A. WELTGE MD

Quality From The Top

Quality From The Top

Working with Hospital Governing Boards to Assure Quality Care

A Guide for Quality Assurance Professionals

James E. Orlikoff

Pluribus Press, Inc., Chicago

© 1990 by Pluribus Press, Inc., and National Association of Quality Assurance Professionals
All Rights Reserved

Except for appropriate use in critical reviews or works of scholarship, the reproduction or use of this work in any form or by any electronic, mechanical or other means now known or hereafter invented, including photocopying and recording, and in any information storage and retrieval system is forbidden without the written permission of the publisher.

94   93   92   91   90   5   4   3   2   1

Library of Congress Catalog Card Number: 90-61321

International Standard Book Number: 0-944496-15-6

Pluribus Press, Inc.
160 East Illinois Street
Chicago, Illinois 60611

National Association of Quality Assurance Professionals
104 Wilmot Road
Deerfield, IL 60015

Printed in the United States of America

Contents

1. Introduction   1
2. Introduction to Hospital Governance: What QA Professionals Need to Know About Hospital Boards   7
3. Hospital Boards and Quality of Care   19
4. What Hospital Boards Should Know About Quality   41
5. Developing Meaningful Board Involvement in Quality   79
6. Providing the Board with Meaningful Quality-Related Information   111
7. Quality and Board Structure   149
8. The Future of QA Professionals   155

Appendix A   159

Appendix B   163

Index   169

About the Author

JAMES E. ORLIKOFF IS PRESIDENT of Orlikoff & Associates, a consulting firm specializing in health care leadership, quality, and risk management. He was formerly the Director of the American Hospital Association's Division of Hospital Governance, and Director of the Institute on Quality of Care and Patterns of Practice of the AHA's Hospital Research and Educational Trust.

Mr. Orlikoff has been involved in quality, leadership, and risk management issues for over ten years. He has designed and implemented hospital quality assurance and risk management programs in four countries, and since 1985 has worked with hospital governing boards to strengthen their oversight of quality assurance and medical staff credentialing. He has authored four books and over forty articles and currently serves on hospital and civic boards. He is the primary author of *Malpractice Prevention and Liability Control for Hospitals.*

CHAPTER ONE

Introduction

When I make presentations to large groups of trustees on the subject of quality of patient care I introduce the topic by asking them a trick question. "How many of you govern a hospital that delivers good quality patient care?" In response, almost all the trustees raise their hands. Then I ask a simple but revealing follow-up question: "How do you *know* that your hospital delivers good quality care?" Most of the hands quickly go down.

When the few intrepid trustees who left their hands up in response to the second question are called upon to state how they know their hospital provides quality care, they tend to respond as follows: "Because we have a quality assurance program"; "Because we are accredited by the Joint Commission"; or "Because we have a Board Quality Assurance Committee."

What these responses to two seemingly simple questions indicate about boards and quality is significant. First, and very importantly, almost without exception every hospital trustee honestly and correctly *wants* to believe that his or her hospital does in fact provide high-quality patient care. Second, few hospital trustees or boards have any real objective knowledge about the quality of care in their hospitals: whether it is bad, ordinary, or good; how it is measured and monitored; and, whether it is getting better or worse. Further, most hospital boards would not understand how to respond appropriately to such information about quality even if they received it. Most boards are unaware of what their role in assuring quality care is or how to discharge it effectively.

Yet, the pressure on hospital boards to accept, refine, and effectively discharge their role in and responsibility for assuring quality care is significant. It continues to grow. The pressure comes from the courts, state legislatures, the federal government, the Joint Commission on Accreditation of Healthcare Organizations, regulatory agencies, the media, the hospital's community, and other sources including some within the hospital and board itself.

Above and beyond these pressures, most hospital governing boards are composed of trustees who sincerely *want* to influence positively the quality of care their hospital provides, who want to do all they can to meaningfully assure a level of quality for all patients that is appropriate to the hospital's mission and resources.

Unfortunately, most trustees and boards do not know *how* to oversee meaningfully and appropriately and work to improve their hospital's quality of care. In the area of quality, hospital governing boards need help.

Hospital quality assurance (QA) and quality management professionals, on the other hand, are in the front-line trenches of the battle for quality. The QA professional knows how to monitor and evaluate quality, has an understanding of what levels of quality the hospital is providing, and knows what action could be taken to improve the hospital's quality of care.

Unfortunately, QA professionals are not always able to act effectively on this knowledge as they are also confronted with considerable pressures and challenges. Regulation and legislation affecting what QA professionals must do—and how to do it—is increasing. The technical knowledge necessary to conduct QA is becoming increasingly complex and demanding. Computerization has created more work for many QA professionals instead of making their job easier. QA professionals are often caught in the middle of extremely stressful political situations within the hospital that make their jobs difficult to perform effectively and that occasionally threaten their very employment.

Moreover, many QA professionals hear about their hospital's "commitment to quality" but see no real support or resources devoted to their activities. Most hospitals have a Chief Financial Officer or someone in a senior position performing the financial function, but very few hospitals have a "Chief Quality Officer" or equivalent in a high management position. Frequently, QA professionals have no time for anything other than generating the paper required for regulatory and accreditation surveys or for supporting the most basic QA functions, let alone actually improving the quality of care.

Many QA professionals are frustrated by these mounting pressures and are becoming skeptical about QA, their hospital's commitment to it, and their own professional future. It would seem that, like boards, QA professionals also need help.

This book proposes that if QA professionals help their boards, they will in turn receive help and support back from their boards. By becoming an internal educator and consultant to their boards, QA professionals can facilitate their board's understanding of its role and responsibility for quality. They can help their board to discharge effectively its quality responsibility and perform its role in overseeing the hospital's quality of care.

The QA professional that works effectively with the board can be instrumental in creating a sincere commitment to quality at the top of the organization. When this happens, and hospital board members and top executives become quality and QA enthusiasts, they send a powerful and compelling message throughout the hospital. As a result, quality and QA becomes important to the organization. Those involved directly in it will have organizational power and prestige conferred upon them.

To achieve this, however, QA professionals must learn about hospital governance and their board. They must learn the proper role of the board in quality and QA, and must address the issue of quality with the board from the perspective of the board. QA professionals must learn to construct and provide meaningful and understandable quality reports to the board; to provide the board with governance information, not management or clinical information. QA professionals must learn to navigate the tricky political seas in the hospital to facilitate an effective team quality effort by the board, medical staff, and management.

In short, the QA professional must learn to gain access to their board and to become an effective educator and consultant on quality to the board. The purpose of this book is to help QA professionals do exactly that.

There are currently many forces in conflict about how quality will be addressed by hospitals in the future, but the issue is as yet unresolved. Consequently, the time is right for most QA professionals to position themselves to improve their board's involvement in quality and thus help improve the quality of care and enhance the visibility and importance of quality and the QA professional in the organization.

While it may not be an easy task for some, the alternatives are indeed worrisome. They include hospitals that have no organizational commitment or integrated approach to quality, or those that are quality-motivated for the wrong reasons. In either of these cases, QA professionals will in all probability spend their days crunching numbers, churning data, and generating paper to meet external requirements instead of spearheading an internally motivated quest for quality in the mainstream of the organization.

This book will introduce QA professionals to the field of governance and how and why it has so dramatically changed. It will review the past and present role of hospital boards in quality, QA, and medical staff oversight. QA professionals will be provided with information that their boards need to know about quality and QA, and with methods and formats for presenting that information to the board. The book will outline techniques for QA professionals to assess their board's involvement in quality and to develop effective board involvement in quality and QA.

By working effectively with their boards, QA professionals can help achieve a sincere top-level, hospital-wide commitment to quality. This will result in improved quality of patient care as well as improved visibility, support, and authority for the QA professional.

CHAPTER TWO

Introduction to Hospital Governance: What QA Professionals Need to Know About Hospital Boards

INTRODUCTION

The board of trustees is a unique group within the hospital. It bears the ultimate responsibility for decision-making and the setting of internal policy, for the financial survival and strategic direction of the hospital, and for the quality of care provided by the institution and all health care providers associated with it.

Yet, in the face of this awesome responsibility it is interesting to note that the majority of trustees who compose hospital governing boards do not possess a high degree of expertise in hospital, health care, or governance issues. The majority of hospital trustees have no formal education in health care management, medicine, nursing, or other areas directly related to health care and hospitals. Further, the vast majority of trustees have never worked professionally within the hospital or health care environment.

One result of this is that the language and acronyms of health care and medicine, the knowledge of regulatory requirements and hospital organization, the understanding of the complex professional and organizational interactions involved in the delivery of care, all of these crucial variables which are second nature to hospital-based professionals are largely foreign or vague to the majority of trustees who govern the hospital.

This significantly relates to how hospital professionals communicate with their boards to facilitate board understanding and decisions regarding complex issues. In order for QA professionals to communicate and interact effectively with their boards regarding quality and QA issues, they must first understand as much as possible about the field of hospital governance generally, and about their board specifically. This understanding will help the QA professional structure the complexity, format, and flow of QA information to the board in such a way as to maximize the board's understanding and involvement in quality and QA. The result is an increase in the board's effective oversight of the hospital's quality and QA program, as well as an enhancement of the board's support of the hospital QA program and reliance on QA professionals. This chapter will present an overview of the evolution of hospital governance and outline current and future issues confronting hospital boards.

The Development and Evolution of Hospital Governance

The governance of hospitals developed with the founding of the nation's first hospitals and has changed through the years as the mission, structure, and activities of hospitals have themselves changed.

The nation's first hospital was founded in 1752 by the nation's first hospital trustee, Benjamin Franklin. Franklin helped establish the Pennsylvania Hospital in Philadelphia by organizing other Philadelphians to donate money for the construction and operation of the hospital. In doing so, Benjamin Franklin established the first developmental stage of hospital governance: the hospital governing board as fund raiser.

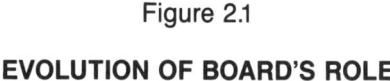

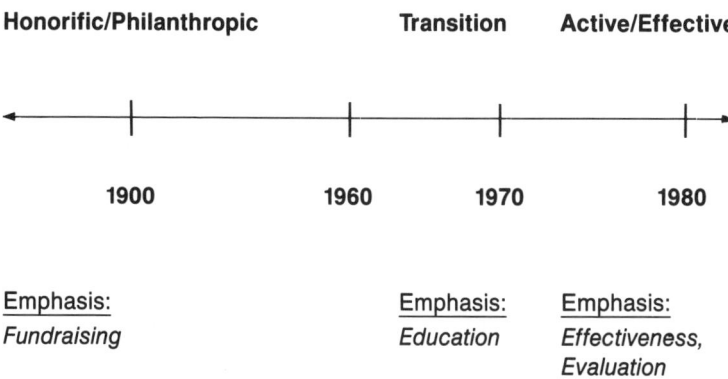

From Benjamin Franklin's beginnings, the role of hospital governing boards evolved in three stages (see figure 2.1). These stages are (Orlikoff, Totten, Ewell):

1. The honorific/philanthropic stage—emphasis on fund raising.

2. The transition stage—emphasis on education.

3. The active/effective stage—emphasis on board effectiveness and self-evaluation.

The first hospital governance stage, the honorific/philanthropic stage, began in 1752 and extended into the mid 1960s. During this stage hospital trustees were selected largely for their fund-raising abilities and for their prestige as community or business leaders. Hospital trustees required little actual knowledge of the hospital and a very slight contribution of time. In fact, early in this period, if a hospital had a deficit at the end of its fiscal year the board of trustees might dip equally into their own pockets to cover the loss. All other hospital-related responsibilities, such as quality of care, were assumed by the medical staff and hospital management.

The second stage of hospital governance, the transition stage, began in the 1960s as a result of the evolution of hospitals from storage facilities for those afflicted with chronic infectious diseases to centers of increasingly high-tech interventive treatment of patients with all forms of acute and chronic illness, as well as for natural processes such as birth and death. As the mission, structure, and function of hospitals changed, new environmental pressures upon hospitals developed. Many of these pressures grew from the advent of government payment for hospital care through Medicare and Medicaid and the resulting trend toward government regulation of hospitals. Other pressures came from the expanding legal liability of the hospital beginning with the landmark case of *Darling v. Charleston Community Memorial Hospital* and the tightening of state hospital licensure statutes that elevated the duties and responsibilities of the hospital board.

As hospitals became more complex and the environmental pressures upon them expanded, the honorific/philanthropic model of hospital governance began to lose its viability. The recognition slowly came to the hospital field that more complex hospitals existing in an increasingly challenging environment required a different model of hospital governance to help preserve the hospital's financial viability and strengthen its strategic direction. Although there was no consensus on what the role of hospital boards should be, there was some consensus that the honorific/philanthropic board was now outmoded, ineffective, and a potential liability to the hospital.

The transition stage was marked by the advent of education for boards and trustees. The concept that motivated this educational trend, and that persists today, is that effective governance of hospitals is directly proportional to the degree of board education regarding their responsibilities and critical health care issues. One direct result of the transition stage was that hospital boards began for the first time to actively govern their hospitals, not simply raise funds for them.

The transition stage led to the third and current model of hospital governance: the active/effective stage. The environmental stage of hospital governance demands boards that understand and effectively discharge their significant roles and responsibilities and satisfy their increasing legal and regulatory accountabilities. The active/effective hospital board also is characterized by an emphasis on continuing education to keep abreast of new developments in the health care field and in the area of governance.

The active/effective board is likely to conduct routine and sincere self-evaluations of itself in an attempt to strengthen its performance and, in so doing, strengthen and improve the hospital it governs. This leads to perhaps the most important characteristic of the active/effective stage of hospital governance, and the one most different from the honorific/philanthropic stage: the concept that the performance of the hospital is directly influenced by the performance of the board.

Not all hospital boards currently function in the active/effective mode. Many boards are still in some degree of transition, while others function in the pre-1960s honorific/philanthropic mode. This is because the three stages of hospital board evolution describe both environmental change as well as individual board change and mode of function. For example, some boards evolved through the stages of governance in response to environmental change. A select few hospital boards evolved in advance of environmental change, perhaps because they anticipated the coming changes. Other boards were so insulated from the environment, so entrenched in the honorific/philanthropic mode that they did not evolve in response to the environment, or indeed at all. Finally, some hospital boards evolved on their own, independent of environmental influence.

The result of this is that the field of hospital governance

is currently a very disparate one. American hospital boards currently cover the wide spectrum of the stages of governance. Furthermore, as the roles and responsibilities of the hospital board become more numerous and clearly defined it is possible to see one board functioning at various levels of sophistication, depending upon the role being performed or the responsibility being discharged. So, a board may function overall in the active/effective mode, but may discharge its quality oversight responsibility in the honorific/philanthropic mode.

It is rare to find a board that performs all of its roles and discharges all of its responsibilities with equal degrees of active/effective sophistication.

The board that functions in the honorific/philanthropic mode, however, tends to be nonadaptive and usually inhibits its hospital from effectively responding to environmental change. When confronted with crisis or challenging circumstances, a board that functions at some level in the honorific/philanthropic mode is far less likely to govern its hospital effectively than is a board functioning in the active/effective mode.

It is important for the QA professional to determine where the board falls generally in the three-mode spectrum of governance function, and to specifically determine how effectively the board discharges its quality oversight and medical staff credentialing responsibility. These assessments are critical for the QA professional to work effectively with their board, and in many cases to structure a program to facilitate their board's evolution from the honorific/philanthropic mode of functioning regarding quality to the active/effective mode. At the conclusion of this chapter guidelines are presented to help the QA professional make a diagnostic assessment of their hospital board's overall mode of functioning. Later in this book, specific guidelines are presented to assess the level of board function in the areas of quality, QA, and medical staff credentialing.

The Roles and Responsibilities of the Hospital Board

As the mission, structure, and function of hospitals changed, the financial, regulatory, liability, and human resource pressures upon hospitals increased. Both the changing role of the hospital as well as increasing environmental pressures upon hospitals combined to redefine and expand the roles, functions, and responsibilities of the hospital board.

For example, during the expansion of hospitals beginning in 1946 with the Hill-Burton Act, money was plentiful and financial solvency was not an issue for the vast majority of hospitals. So, during this period, close financial oversight of the hospital was not an expected or critically necessary function of most hospital boards. As dollars became scarce, and financial pressures upon hospitals mounted, finance and the financial viability of the hospital became a major focus of most hospital boards, often to the exclusion of other key areas—such as quality.

Different individuals, organizations, state legislatures, regulatory agencies and hospitals and multi-hospital systems have different views of what exactly the roles, responsibilities, and functions of a hospital board should be. The fact that there is little consensus is a significant reflection of the great disparities in the field of hospital governance and helps explain why more boards do not have a clear definition of their own purpose and function. A detailed guideline, from the American Hospital Association, on the leadership responsibilities and functions of a hospital board appears in Appendix A.

Assessing a Board's General Mode of Functioning

As indicated earlier, the three evolutionary stages of hospital governance can be used as a framework to describe general modes of how a board functions. Once a QA professional has learned about the evolution of hospital governance, the first step to work effectively with the board is to assess the board's position in the general three-mode spectrum of governance. Determining what mode the board is generally in will enable the QA professional to design a beginning strategy for facilitating effective board oversight of quality and for developing a productive working relationship with the board.

In making such an assessment it is tempting to primarily use structural variables of the board such as board size, number of yearly meetings, number of committees, length of terms of office and the like to measure board sophistication. This should be avoided. Although board structure is easier to measure than board function, board structure does not necessarily predict board function. Cumbersome board structures do not preclude effective board function, though they may inhibit it. Conversely, streamlined board structures do not insure effective board function, though they may facilitate it. (The results of an American Hospital Association survey on the characteristics of hospital boards are contained in Appendix B.)

For example, although several governance consultants advise that a smaller board will function more effectively than a larger one, it is not uncommon to find boards of over twenty-five members functioning effectively and boards of twelve members or less hardly functioning at all. Certain structural variables occurring together, however, such as an extremely large board that meets only twice a year without an executive committee that meets in the interim, may well indicate a board operating in the honorific/philanthropic mode.

Even though such combinations of structural variables may indicate the mode of board function, the best measures of board function are those that indicate how well the board defines, performs, and evaluates its roles and responsibilities. Ultimately, the best measure of board function is the degree to which it positively influences the functioning and viability of the institution it governs.

Boards functioning in one of the three general modes

of governance often exhibit certain characteristics. It is important to note that no single characteristic defines a board's overall mode of functioning. In fact, one board may exhibit characteristics from all three modes. Also important is to realize there are relative degrees of board sophistication within each mode. The more characteristics of a particular mode exhibited by a board, the more entrenched the board is in that mode.

Characteristics of Boards by Functional Mode

1. *The Honorific/Philanthropic Board*

No defined statement of roles and responsibilities.

No regular evaluation of hospital mission.

No development or evaluation of long-range strategic plan.

No financial targets, or vague targets.

No actual, detailed knowledge of hospital's financial condition.

No formal performance objectives for CEO, loose or nonexistent performance appraisal of CEO.

"Rubber Stamps" medical staff credentialing decisions.

No written charge or role definition for board committees.

No annual workplans for board or board committees.

No formal orientation program for new trustees.

No selection criteria for new trustees.

No continuing education activities for individual trustees or board as a whole.

No conflict of interest or board confidentiality policies.

No board self-evaluations conducted.

No oversight of quality, or QA.

No limit on trustee terms of office.

Board has not agreed upon, or has ill-defined, role in relating to the community.

2. *The Transition Board*

Has begun to investigate and agree upon what the roles, responsibilities, and functions of the board should be.

Has recently revised, or is currently examining the structure and composition of the board.

Has begun to examine the board-CEO relationship and board-medical staff relationship.

Has recently initiated an orientation program for new trustees.

Has for the first time sent several trustees to outside education programs on hospital governance or health care issues.

Has conducted continuing education programs for entire board.

Is planning, or has conducted, first board self-evaluation.

Is becoming concerned about its role in assuring quality.

Is examining its relationship to the community.

3. *The Active/Effective Board*

Has a written statement of board roles, responsibilities, and functions.

Evaluates institutional mission at regular intervals.

Oversees development, implementation, and monitoring of strategic plan.

Has financial targets for the institution.

Has detailed knowledge of hospital's financial condition.

Has goals and objectives for CEO performance, conducts formal and routine CEO evaluations, may have employment contract with CEO.

Does not "rubber stamp" medical staff credentialing recommendations, rather evaluates recommendations against criteria in bylaws.

Has written charge and definition of role for committees of the board.

Develops and approves annual workplan for board and for board committees.

Has formal orientation program for new trustees.

Encourages trustees to attend external education programs and reimburses their expenses.

Conducts regular board continuing education programs.

Conducts regular and sincere board self-evaluations which result in action plans that are followed to improve board performance.

Works to continually clarify the relative roles, responsibilities, and functions of the CEO, medical staff, and board.

Works to maintain good board-medical staff relations.

Receives regular quality-related information.

Defines the difference between management information and governance informations and requires governance information.

Involves its key trustees in legislative activities.

Limits trustee's term of office—thus helping to prevent the board from becoming too inbred and "business as usual" oriented while continually introducing new skills and perspectives.

Believes that the performance of the board directly influences the performance of the hospital and acts to discharge its roles and responsibilities effectively and efficiently.

Has defined its role in relating to the community.

Conclusion

To work effectively with the hospital board the QA professional must first attempt to learn as much as possible about the general field of hospital governance. Then, using this knowledge as a foundation, the QA professional should learn as much as possible about the characteristics and general mode of functioning of their hospital board.

This chapter has outlined the general field of hospital governance, with an emphasis upon the evolution of governance and the resulting changing roles and responsibilities of the hospital board. It has also provided a framework for the QA professional to assess their board's general mode of functioning. This assessment will then allow the QA professional to communicate effectively with the board and to begin to tailor a board-specific strategy for improving the board's involvement in and oversight of quality and QA. A more detailed instrument for assessing the board's current role in quality and QA is presented in chapter 5.

References

1. *Darling v. Charleston Community Memorial Hospital,* 33 Ill2d 326, 211 NE2d 253 (1965).
2. Orlikoff J, Totten M, Ewell C. *Governance Reference Manual.* Work in progress.

CHAPTER THREE

Hospital Boards and Quality of Care

THE HISTORICAL ROLE OF THE HOSPITAL BOARD REGARDING QUALITY

Historically, before hospitals or hospital boards existed, the responsibility for the provision of quality care rested solely with the individual who provided the care: the physician. The accountability for that care also rested with the individual physician, as there was no organization or institution for the physician to affiliate with or practice within. In one of the very first laws that addressed the accountability and liability for the provision of substandard care, the physician was held to be ultimately accountable. More than 1,600 years prior to the Oath of Hippocrates, the Code of Hammurabi (about 2,000 B.C.) stated:

> If the surgeon has made a deep incision in the body of a free man and has caused the man's death or has opened the caruncle in the eye and so destroys the man's eye, they shall cut off his forehand (Kramer, 1976).

From this starting point both the responsibility and accountability for the quality of care rested solely with the physician and only gradually over the centuries began to expand to include the institutions within which physicians practiced: hospitals. Even this initial institutional responsibility rested exclusively with the organized physicians within the hospital: the medical staff. It was only during the twentieth century, and specifically during the late 1950s and 1960s—the transition stage, that the responsibility and accountability for the quality of care provided by a physician practicing within a hospital, as well as the care and services provided by the hospital itself, began to expand to include the corporate institution itself and the board that governed it.

From the time of the first American hospital until the transition period, the hospital was regarded as simply a facility within which physicians and surgeons were allowed to admit and treat their patients. This perspective along with the protection from liability that hospitals enjoyed due to the doctrine of charitable immunity resulted in a conceptual and organizational separation between the management and medical staff of the hospital. Southwick (1973) states that "The role of hospital management was limited to purely financial and housekeeping chores. The practice of medicine was solely a matter for the medical staff physicians." If management's role was so limited, what was the role of the hospital board?

During the honorific/philanthropic stage, with the vast majority of hospital boards operating in that mode, the role of the hospital board regarding oversight of the medical staff was next to nothing and its role in overseeing quality was nonexistent. The responsibility for assuring a qualified and skilled medical staff was primarily delegated to the medical staff. The management's responsibility was to provide a competent nursing, support, and maintenance staff. The legal and conceptual environment required no board responsibility and accountability for quality and so trustees believed their boards and them had no quality responsibility and acted accordingly.

Figure 3.1, Caniff (1980), graphically depicts the legal and conceptual relationship between the hospital governing board, administration, and medical staff as it existed prior to

Figure 3.1

Legal Relationships among the Governing Body, Administration, and Medical Staff in the Past

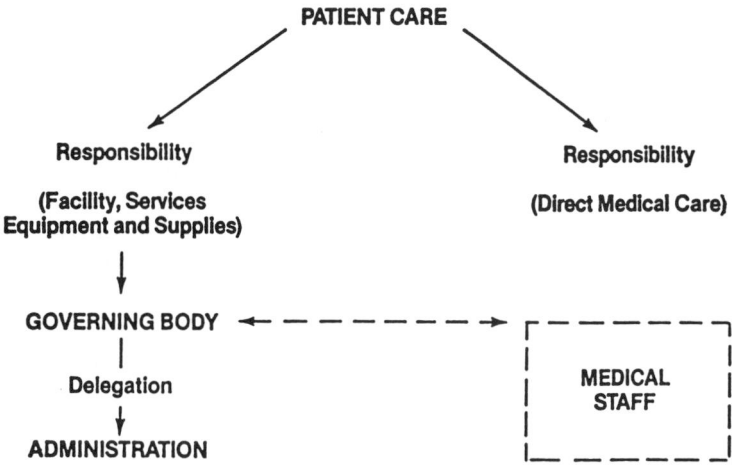

the transition stage. Notice that the exclusive responsibility of the board was the facility itself, the equipment, nonmedical services, and supplies. The board then delegated this responsibility to administration. Notice also, that figure 3.1 expresses that the responsibility for both the provision of direct medical care and the assurance of the quality of that care was solely the responsibility of the medical staff. The board did not delegate this responsibility to the medical staff as it was not the board's responsibility to delegate. Further, as figure 3.1 indicates, the board did not have organizational authority as such over the medical staff and there was, at best, only a tangential relationship between them.

If during the early twentieth century a hospital board felt it should more actively oversee the activities of the medical staff or question its credentialing recommendations, the medical staff and the courts would put the board squarely back in its place. Such was the situation in the 1902 Michigan Supreme Court ruling in the case of *Pepke v. Grace,* which spoke pointedly to the role of the hospital board in medical staff credentialing:

> The trustees of a hospital are laymen. The rules of the hospital provide for a medical board... who have charge of all surgical matters in the hospital. They examine applicants... and recommend such appointments to the trustees.... The trustees, who are laymen, must naturally leave the competency of the physician... to the judgement of those competent to determine such matters, since they are not qualified to make the determinations themselves. [The board] performed their full duty... in appointing a [medical] board to examine applicants.

The *Pepke v. Grace* ruling contributed to the persistent perspective that the board's role in deciding on medical staff admissions and on delineation of clinical privileges was little more than simply approving without review or question all recommendations from the medical staff credentials or executive committee. Unfortunately, this "rubber stamp" perspective of the board's role in medical staff credentialing can still be seen in many boards today.

Only twelve years after *Pepke v. Grace,* another court ruling soundly emphasized the hospital governing board's complete lack of responsibility for the quality of medical care provided by the hospital. In *Schloendorff v. Society of New York Hospital* (1914) the New York State Court of Appeals held that hospitals were only responsible for the ministerial care or administrative acts of their employees. Further, the court ruled that hospitals, and hence their boards, were not responsible for the medical care provided by physicians or nurses, or other health care providers who, like nurses, were regarded by the court as operating under the direction of the physician.

The *Pepke v. Grace* and *Schloendorff* rulings clearly demonstrate the environmental influences that pressured boards to function in the honorific/philanthropic mode, especially in the areas of medical staff credentialing and oversight of quality. The majority of hospital trustees believed, consistent with environmental pressures, that they were basically hospital custodians and had no responsibility for sincere oversight of the medical staff or for the quality of care.

In the late 1950s, however, the environment began to change. Certain malpractice cases began to redefine the hospital's responsibility for overseeing the medical staff and assuring quality care as well as to expand the hospital's liability for failing to do so. These cases marked the beginning of the transition stage in hospital governance as the increasing responsibilities of the hospital for quality and credentialing gradually expanded to include the ultimate authority of the hospital—its governing board.

The first stirrings of the transition stage began when the New York Court of Appeals ruled that the doctrine of charitable immunity no longer applied to hospitals in the case of *Bing v. Thunig* (1957). A statement of the court in that case demonstrates the expansion of the responsibility and liability for quality beyond the individual physician or organized medical staff to include the hospital, and portends the future responsibility of the board.

> The conception that the hospital does not undertake to treat the patient, does not undertake to act through its doctors and nurses, but undertakes instead simply to procure them to act upon their own responsibility, no longer reflects the fact. Present day hospitals...do far more than furnish facilities for treatment. They regularly employ on a salary basis a large staff of physicians, nurses, and interns...and they charge patients for medical care and treatment, collecting for such services, if necessary, by legal action. Certainly, the person who avails himself of "hospital facilities" expects that the hospital will attempt to cure him, not that its nurses or other employees will act on their own responsibility. Hospitals should, in short, shoulder the responsibilities borne by everybody else.

The wheels of motion of the transition stage were accelerated by the landmark *Darling v. Charleston Community Memorial Hospital* (1965) malpractice case. Interestingly, although the Darling case and its implications are well known to most, if not all, QA professionals, hospital risk managers, and CEOs, few trustees are aware of the case or its significance. The *Darling* case stimulated additional environmental change, such as state hospital licensing statutes and Joint Commission quality assurance standards, which in turn heightened the board's responsibility for quality and oversight of the medical staff.

This can be seen in the following language quoted from the first hospital licensing statute enacted in the state of Michigan, the *Pepke v. Grace* case, by the Michigan Legislature in 1968 (Kitch, 1986):

> The governing body of each hospital shall be responsible for . . . the selection of the medical staff, and the quality of care rendered in the hospital. The governing body shall . . . insure that physicians admitted to practice in the hospital are granted hospital privileges consistent with their individual training, experience and other qualifications; and insure that physicians admitted to practice in the hospital are organized into a medical staff in such a manner as to effectively review the professional practices of the hospital for the purposes of reducing morbidity and mortality. . . .

When contrasted to the language of the *Pepke v. Grace* case quoted earlier, it is easy to see the 180 degree revolution that occurred in the board's responsibility for quality and credentialing. This and other significant environmental changes were supposed to reverse hundreds of years of board apathy and inaction on quality and credentialing matters. But the habits of history were not surrendered by boards easily.

Regardless of the willingness or abilities of boards to accept and adjust to their new and growing responsibilities for quality and oversight of the medical staff, the environ-

Figure 3.2

Legal Relationships among the Governing Body, Administration, and Medical Staff at the Present Time

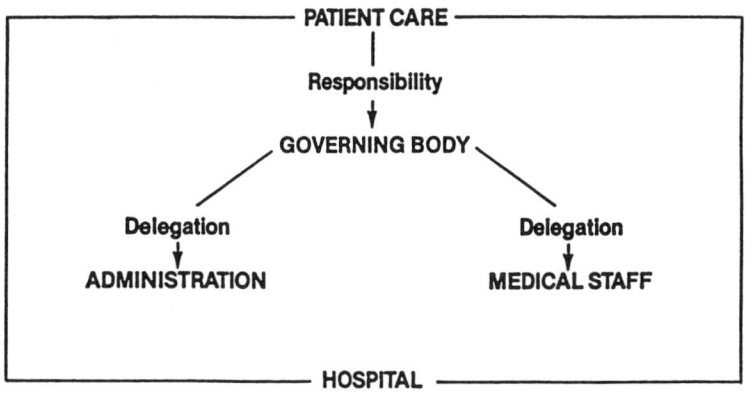

ment had changed forever. As Caniff (1980) demonstrates in figure 3.2, the results of the *Bing v. Thunig* and *Darling* decisions and the changes in state laws and regulatory requirements they stimulated were to dramatically alter the legal, conceptual, and organizational relationship between the hospital governing board, medical staff, and management. Notice that figure 3.2 places the responsibility for quality patient care, and everything within the hospital, squarely on the shoulders of the board. The board divides and delegates this responsibility to the administration and medical staff, but retains ultimate responsibility. Further, the board is the ultimate authority of the hospital, with organizational authority over the medical staff. Quite a difference from the board-medical staff relationship depicted in figure 3.1.

Figure 3.2 demonstrates the organizational structure of the hospital from the legal and conceptual perspective from the immediate post-*Darling* period to present day. It does not reflect, however, the actual practice of many, if not most, boards relative to quality and medical staff credentialing.

The Current Role and Responsibility of the Board Regarding Quality

The hospital environment that once encouraged boards to function in the honorific/philanthropic mode now requires that they operate in the active/effective mode regarding quality and medical staff credentialing. This can be seen in the standards of the Joint Commission on Accreditation of Healthcare Organizations, the Medicare participation requirements, all fifty state hospital licensure laws, and in volumes of case law.

The standards of the Joint Commission on Accreditation of Healthcare Organizations (JCAHO) specifically emphasize the board's responsibility for quality. This is evident throughout the JCAHO Governing Body standards and required characteristics of the Accreditation Manual for Hospitals (JCAHO, 1990):

> GB.1 An organized governing body, or designated persons so functioning, is responsible for establishing policy, *maintaining quality patient care* (emphasis added), and providing for institutional management and planning.
>
> GB.1.11 The medical staff executive committee makes recommendations *directly to the governing body for its approval* (emphasis added)... such recommendations pertain to at least the following:...
>
> GB.1.11.1.2 The mechanism used to review credentials and to delineate individual clinical privileges.
>
> GB.1.11.1.3 Individual medical staff membership;
>
> GB.1.11.1.4 Specific clinical privileges for each eligible individual;
>
> GB.1.11.1.5 The organization of the quality assurance activities of the medical staff as well as the mechanism used to conduct, evaluate, and revise such activities;...

Other standards and required characteristics in the

Governing Body section of the JCAHO manual stress both overall and specific board responsibilities for quality:

> GB.1.15 The governing body requires a process... designed to assure that all individuals who provide patient care services, but who are not subject to the medical staff privilege delineation process, are competent to provide such services....
>
> GB.1.16 The governing body requires mechanisms to assure the provision of one level of patient care in the hospital....
>
> GB.1.18 The governing body provides for resources and support systems for the quality assurance functions and risk management functions related to patient care and safety....
>
> GB.3.2 All members of the governing body are provided information relating to the governing body's responsibility for quality care and the hospital's quality assurance program.

The JCAHO standards and required characteristics relating to board responsibility for quality and credentialing are quite clear and specific. Yet, the recency of the emphasis on the board's responsibility for quality is evident in the fact that the "Governing Body" chapter of standards became effective for accreditation purposes on April 1, 1984. Further, the requirement that the board support the quality assurance and risk management functions (GB.1.18 quoted above) only became effective for accreditation purposes on Jan. 1, 1989.

In addition to its recency of implementation, the JCAHO required characteristic GB.1.18 is significant in terms of what it requires regarding board support for QA and risk management. The JCAHO draft scoring guidelines for GB.1.18 are quite specific and, in part, state (JCAHO, 1988):

Elements of Satisfactory Performance

1. The governing body requires performance of and has delegated responsibility for the following quality assurance/risk management functions:

- Identification of general clinical areas that represent actual or potential sources of patient injury

- Development and use of an indicator based approach to identify and evaluate individual cases of undesirable or adverse patient-care occurrences within the general clinical areas

- Resolution of clinical problems disclosed through data evaluation

- Provision of education for all staff on approaches to reducing clinical sources of patient injury

2. The governing body receives reports at least every six months on:

- The frequency, severity, and causes of undesirable or adverse patient-care occurrences; and

- Actions taken and the results of actions taken to reduce the occurrences' frequency and severity or to eliminate their causes.

The Conditions of Participation in the Medicare program also require board responsibility for quality. Anthony and Singer (1989) point out that to be eligible for Medicare reimbursement, a hospital must have an effective governing body that is the ultimate legal authority of the hospital; and further, that the board must assure that the medical staff is accountable directly to the board for the quality of care.

Recent legal decisions in malpractice cases have either relied upon or expanded these requirements by increasing the liability of the hospital and the board for poor quality care and ineffective oversight of the medical staff. In 1975, the Supreme Court of New Jersey issued a significant ruling in *Corleto v. Shore Memorial Hospital.* The court held that the trustees of the board and the entire medical staff could be held liable if it could be demonstrated that any of those individuals knew that a physician was incompetent and was at all legally responsible for the granting of the incompetent physician's privileges.

In 1981, the Wisconsin Court of Appeals emphasized the board's responsibility for quality and meaningful oversight of the medical staff in the *Johnson v. Misericordia Community Hospital* malpractice case. In this case, a Dr. Salinsky claimed on his Misericordia medical staff application that his privileges had never been suspended or revoked at another hospital, which was untrue. He also failed to answer questions on the application regarding malpractice insurance. Yet the medical staff recommended, and the board approved, his admission to the medical staff. Shortly thereafter, Dr. Salinsky was elected chief of the medical staff and, as such, he endorsed and recommended his own request for orthopedic privileges, which was subsequently approved by the board. Dr. Salinsky, while performing orthopedic surgery, later injured a patient by severing the right femoral artery and nerve.

The Wisconsin Court of Appeals ruled against Misericordia and Dr. Salinsky and in its decision stated:

> [It is] clear that the medical staff is responsible to the governing body of the hospital for the quality of all medical care provided patients.... The [medical] staff must be organized with a proper structure to carry out the role delegated to it by the governing body. All powers of the medical staff flow from the board of trustees, and the staff must be held accountable for its control of quality.... The protection of the public must come from...the Hospital Board of Trustees.

Many other legal cases have broadened the liability of the hospital and the responsibility of the board as well. These cases have expanded the duties and responsibilities of the hospital and board that were first framed in the *Bing v. Thunig* and *Darling* decisions. A net effect of these malpractice liability cases has been to establish and reinforce the ultimate legal responsibility of the hospital board for the quality of care rendered, for medical staff appointments and privilege delineations, and for the routine and systematic evaluation of all health care providers. The responsibility of the board, however, is not simply limited to monitoring the quality of care and activities of the medical staff. The board must act to correct situations of poor quality or physician incompetence, or risk exposing the hospital, and the board itself, to liability.

Thus, the current role and responsibility of the hospital board regarding quality includes continuous monitoring and analysis of the quality of care and activities of the medical staff, but is not limited to this. The board must also act when necessary to correct quality problems and suspend or revoke a medical staff member's privileges when appropriate. Further, the board must emphasize quality as a hospital priority, insure that sufficient resources are devoted to support QA and related activities, and lead the hospital to an accepted definition of quality and in the constant pursuit of excellence in quality care.

These are very significant and relatively recent board responsibilities. That the board is the ultimate responsible authority for quality is an accepted principle and legal fact. That the vast majority of hospital trustees are uncomfortable with this responsibility and unclear how to effectively discharge it is also a fact.

Why Many Boards and Trustees Are Uncomfortable with Their Responsibility for Quality

Although the board's responsibility for quality is clear and its authority over both management and medical staff is beyond debate, the majority of individual trustees are uncomfortable with this responsibility and authority. Consequently, many hospital governing boards know next to nothing about the quality of patient care provided in their hospital or the clinical competence of the medical staff. Further, these boards have neither the inclination nor the tools to properly monitor quality or the integrity of the medical staff credentialing process.

These assertions are reinforced by recent governance research and by the author's consulting experience with hospital boards. The Hospital Research and Educational Trust survey on governance (Alexander, 1986) revealed the top ten agenda items that occupied most of a hospital board's time in 1985. The relative importance that boards attached to different agenda items and issues is reflected by the amount of time that boards devoted to each issue. The topic of quality as such was, remarkably, not one of the top ten agenda items that boards focused on in 1985.

Agenda Issues Occupying Most of a Board's Time

1. Financial Viability . . . . 29.4 percent

2. Strategic Planning . . . . 18.4 percent

3. Diversification/Mergers/Joint Ventures . . . . 14.0 percent

4. Capital Projects . . . . 12.0 percent

5. Competitive Position . . . . 11.0 percent

6. Medical Staff Appointments . . . . 5.5 percent

7. Other . . . . 3.3 percent

8. Patient Care Standards . . . . 2.7 percent

9. Litigation . . . . 1.9 percent

10. CEO Performance . . . . 1.3 percent

The three quality-related items that were included in the top ten agenda list were given extraordinarily small amounts of board attention. The percentage of time devoted to the combined agenda items of medical staff appointments, patient care standards, and litigation was approximately one-third of the time devoted by boards to finance.

More recent research reaffirms this tendency but also indicates that more boards are becoming conscious of the critical importance of quality. A 1989 survey conducted by Korn/Ferry and the University of Minnesota at Minneapolis (*Hospitals,* 1989) of 750 general acute care hospital CEOs with a response rate of 45 percent found the following issues were rated highest in terms of dominating board agendas:

1. Improving profitability — 80.1 percent
2. Strategic planning — 65.1 percent
3. Physician relations — 40.1 percent
4. Hospitalwide quality improvement — 38.9 percent

Even though quality is ranked as a dominant issue by 38.9 percent of the 350 responding CEOs, it is still dwarfed by financial and planning issues. Commenting on the survey results, a Korn/Ferry senior partner stated "there is little reference to the board's role in quality of care, medical and ethical issues, and the things that are at the heart of the hospital" (*Hospitals,* 1989).

These research findings indicate that most boards regard their primary responsibility as insuring the financial viability of the hospital. While finance is indeed a critical responsibility for boards, so is quality. The problem with boards that define their primary, if not exclusive, responsibility as financial is that they usually assume that if they take care of finance then "somebody else," probably the medical staff, will take care of quality. This assumption creates in the board a conceptual organizational structure of the hospital that is identical to the actual organizational structure common to hospitals in the pre-*Darling,* honorific/philanthropic period demonstrated earlier in figure 2.1

Most trustees are much more comfortable with financial and other routine business issues typical of the focus of non-hospital boards than they are with quality. Many hospital trustees have first served on other business, civic, and nonprofit boards where the primary focus is finance or business or perhaps service, but where there is no issue similar in content or complexity to quality or medical staff oversight.

The common refrain heard from many hospital trustees and felt by many more is "I am not a doctor or a health care professional, therefore, how can I understand quality or presume to oversee the activities of the medical staff?" This view is often reinforced by medical staff members who, either implicitly or explicitly, advance the notion that to understand quality or physician competence one must be a physician and that quality is therefore best left to the medical staff. Consequently, many trustees do not understand quality or the board's significant responsibility for it and are intimidated by the notion of addressing quality. Many trustees are also intimidated by the implied challenge to physicians and resulting conflict they perceive as likely to occur should they address quality issues.

Even those boards that do devote significant amounts of agenda time to quality still may be ineffectively performing their quality oversight role. Further, these boards, regardless of the amount of time they spend on quality, are often composed of trustees who are unclear about or intimidated by quality. A board that simply spends time on "quality" has no guarantee that the time is well spent. Unless a board has been educated about its responsibilities for quality and has agreed upon its role in overseeing quality and medical staff credentialing and, most importantly, has defined the quality-related information it will receive and what questions to ask or actions are appropriate to take in response to that information, the odds are that no matter how much time that board spends on quality—it is wasted time.

The 1983 book *Keys to Better Hospital Governance Through Better Information,* written by Barry Bader, was one of the first to emphasize that the management, medical staff, and board each have different responsibilities and so require different types of information. The quality-related information that is presented to most boards is not tailored to the responsibilities of the board. It may be far too patient-specific, and thus be more appropriate information for medical staff committees than for the board. Or, the information may be far too general and out-of-focus to facilitate meaningful governance involvement and oversight.

To effectively oversee quality and medical staff credentialing, boards need *governance* oriented information about quality, not management or clinically oriented information. Unfortunately, many boards that do spend significant amounts of time on quality do not receive meaningful, understandable, governance oriented quality-related information. In these situations, the trustees are likely to become more uncomfortable than ever about quality. After all, they agreed that quality is important and have been devoting board time to it, but *still* they can not understand it or answer such seemingly simple questions as: Is our hospital providing good quality care? Is the quality of care improving? What can we do to help improve quality? This can obviously be very frustrating and demoralizing for a board.

There are many inherent difficulties and pitfalls in hospital board oversight of quality and the medical staff which make the frustration and lack of comfort felt by many trustees understandable. Bader (1983) points out that:

> These factors make the governance role in assuring the quality of care more difficult and controversial than its role in assuring, say, financial integrity and viability. On the one hand, by over-zealously asserting their legal responsibility, trustees

may begin interfering in patient care decisions and alienating community physicians. On the other hand, if trustees entrust quality assurance entirely to the medical staff, they will actually weaken the medical staff's own quality assurance efforts as well as abrogate the board's legal accountability for quality care.

Such is the delicate balance that the board must strike and maintain to effectively perform its quality and medical staff oversight role. The risks of missing or losing this balance are several and serious. Board under-involvement generally insures an unempowered QA program due to lack of top-level commitment and risks serious quality problems, patient injury, liability, and bad public relations with a subsequent deterioration of the relationship with the community. Conversely, inappropriate board over-involvement can cause or exacerbate serious political conflicts with the medical staff and place the CEO and QA professional in an extremely difficult situation, damage the QA program and also reduce quality of care.

There are many valid reasons why most trustees feel uncomfortable about their responsibility for and role in quality. Among them are: lack of understanding of the extent of their responsibilities and those of the medical staff, management, and the QA professional; no guidance in how to discharge those responsibilities; not understanding what "quality" is; not knowing how to analyze and act upon quality-related reports provided to the board; fear of discovering that their hospital may be delivering poor quality care; fear of conflict with the medical staff and possible litigation and liability; and concern about changing regulatory requirements and case law which expands their quality responsibilities and potential liability even further.

These last two points are worth noting as inhibiting factors to effective board involvement in quality which are likely to increase. For example, at the time of this writing one requirement under consideration for reporting physician disciplinary action or privilege suspension to the National Practitioner Data Bank, established by the Health Care Quality Improvement Act of 1986, is that the reason for the action must be reported directly from the narrative of the minutes of the hospital governing board meeting where the disciplinary action was taken. The chilling effect that this requirement could have on the willingness of boards to take such disciplinary action is considerable. Boards may be fearful of engendering lawsuits by physicians should they approve a recommendation of the medical staff to restrict or revoke a medical staff member's privileges. Much more disconcerting to boards will be the circumstance where they should reverse a medical staff credentialing recommendation or, in a rare situation, initiate such action themselves. Because of the ramifications of having disciplinary and negative credentialing decisions on a physician's National Practitioner Data Bank record for life, some medical staffs will be reluctant to recommend action against physicians and, in essence, pass the buck to the board.

The buck, however, does indeed stop with the board. As regulation and litigation increase, the pressures upon boards to understand and effectively discharge their responsibilities will grow. So, too, will the discomfort level of boards grow unless their responsibilities are made clear to them and unless they are presented with meaningful information that facilitates their understanding of quality issues and their taking appropriate, decisive action.

The Role of the Quality Assurance Professional in Board Oversight of Quality

As the role of the board regarding quality expands, the hospital QA professional will become more involved with the board. This involvement may be limited to preparing written quality reports to the board, may involve staffing the board committee on quality if one exists, or may include attending meetings of the board to present quality-related reports and support board deliberations and decisions on quality issues. Regardless of their current level of involvement with their board, QA professionals should regard themselves as, and structure their activities to become, educators and consultants to their boards.

By working effectively with their boards, QA professionals can facilitate a sincere board commitment to, and meaningful board involvement in and oversight of quality. Furthermore, as boards become more involved in and committed to quality they will become more convinced of the importance of the hospital's QA program and the value of the hospital's QA professionals. Thus, by interacting effectively with their boards, QA professionals can achieve a double benefit: more effective board oversight of quality; and, greater visibility, stature, and power within the hospital for the QA professional.

To become an effective educator and consultant to the board, the QA professional will need to address many substantive, organizational, and political issues within the hospital. The substantive issues include: how to educate the board about its role in quality; how to make the board comfortable with that role; how to turn volumes of QA data into meaningful governance information for board review; how to develop a common QA vocabulary with board, medical staff, and management; how to facilitate agreement on the relative roles of board, medical staff, and management regarding quality; and, how to motivate the board to become quality leaders in the hospital and to communicate the quality message from the top.

The organizational and political issues the QA professional faces include: balancing relationships and loyalties between management, medical staff, and board; gaining access to the board; avoiding being caught in the middle of conflicts between board, medical staff, and management; and elevating the organizational stature of QA without causing damaging political rifts within the management structure.

Other significant issues also exist as the QA professional prepares to become educator and consultant to the board. Perhaps the most critical issue is gaining the approval and support of the hospital CEO to work with the board to enhance its oversight of the hospital's quality and medical staff. This should become an easier task as the environmental pressures upon boards to effectively oversee quality mount and CEOs recognize that they must do all they can to educate and involve their boards in quality. For example, in 1988, the fifth most common hospital contingency citing by accreditation element by the Joint Commission on Accreditation of Health Care Organizations was for lack of compliance with governance standards and the twenty-sixth most common was for shortcomings in board involvement in quality assurance (American Hospital Association, 1989). As these and other pressures increase, a thoughtful QA professional who is familiar with governance and has a plan for educating and involving their board in quality will be a welcome resource to the strategic-minded CEO.

Thus the role of the QA professional is twofold. First, the QA professional must position themself within the hospital to *become* educator and consultant to the board. Next the QA professional must function as and remain an effective educator and consultant to the board. Accomplishing these tasks is critically important for the QA professional to gain the support of the board and to elevate the visibility and importance of quality in the organization. It is also crucial for the board to have an effective educator and consultant to become both comfortable with their role in quality and proficient at discharging it. The remaining chapters of this book focus on helping the QA professional accomplish these challenging but essential tasks.

References

1. Alexander JA. *Current Issues in Hospital Governance.* Hospital Survey Series. Hospital Research and Educational Trust. Chicago, 1986.

2. *Accreditation Manual for Hospitals.* Joint Commission on Accreditation of Healthcare Organizations. Chicago, 1990.
3. American Hospital Association. "Technical Briefing: The Joint Commission Quality Assurance Model." Division of Quality Control Management. AHA, Chicago, February 1989.
4. Anthony MF, Singer LE. "The legal basis of the board's quality assurance duties." *Trustee.* January 1989.
5. Bader BS. *Keys to Better Hospital Governance Through Better Information.* Camp Hill, Pa: Hospital Trustee Association of Pennsylvania, 1983.
6. *Bing v. Thunig,* 2 NY2d 656, 143 NW2d3 163 (N.Y.S. 2d3 1957).
7. Caniff CE Jr. "Responsibilities and relationships of the medical staff, administration, and governing body." Presented at the American Medical Association Seminar "Medical Staff: Physician, Friend, or Foe," Chicago, April 24, 1980.
8. *Corleto v. Shore Memorial Hospital,* 350 A2d 534 (NJ 1975).
9. *Darling v. Charleston Community Memorial Hospital,* 33 Ill2d 326, 211 NE2d 253 (1965), *cert denied* 383 US 946 (1966).
10. "Draft Scoring Guidelines for Hospital Risk Management Activities." Joint Commission on Accreditation of Healthcare Organizations. Chicago, Dec 1988.
11. "Executive advancement—CEOs should spend more time with Boards." *Hospitals,* July 1989.
12. *Johnson v. Misericordia Community Hospital,* 99 Wis2d 708, 301 NW2d 156 (1981).
13. Kitch RA. "Legal Liability in Conjunction with Credentialing." Kitch, Saurbier, Drutchas, Wagner & Kenney, P.C. Detroit, Michigan, November 1986.
14. Kramer C. *Medical Malpractice.* 4th ed. New York City: Practicing Law Institute; 1976.
15. *Pepke v. Grace* 130 Mich 493 (1902) p. 496.
16. *Schloendorff v. Society of New York Hospital,* 211 NY 125, 105 NE 92 (1914).
17. Southwick AF. "The hospital as an institution-expanding responsibilities change its relationship with the staff physician." *California Western Law Review.* 1973. Spring. 9:429.

CHAPTER FOUR

What Hospital Boards Should Know About Quality

INTRODUCTION

The purpose of this chapter is to outline in linear fashion what hospital boards need to know about quality of care, quality assurance, and the medical staff credentialing process. This chapter may be of use to QA professionals as a guide in constructing an orientation and continuing education program for boards. Also, this chapter serves as a reminder to QA professionals that their level of sophistication and technical expertise regarding quality far exceeds that of the trustees on their hospital board. Consequently, the material may seem basic to the QA professional—as well it should. Bear in mind, however, that it is not written to educate QA professionals about quality or QA, but to help QA professionals understand the education process most *boards* need regarding quality and QA.

When interacting or communicating with their boards, QA professionals must keep in mind that their QA knowledge, technical vocabulary, and awareness of the detailed specifics of regulatory requirements and quality monitoring techniques is necessary for them to do their jobs; but, that same detailed level of knowledge is not necessary for the board to do *its* job. The QA professional's role is not to educate their board to be a QA professional; this is, unfortunately, a common mistake. Burying boards in detailed QA matters tends to have one of two negative results: the board becomes inappropriately over-involved in quality and QA; or, the board becomes numb and disinterested at being bombarded with QA information that is management and clinically, but not governance, oriented.

The QA professional must learn to educate the board about quality at a level that is appropriate to the active/effective functional mode of *governance*. Then the QA professional must continue to communicate with the board about quality at a level of information detail that is appropriate to the board. This and the following chapters are designed to help the QA accomplish both tasks.

Why Boards Are Responsible for Quality

There are four broad reasons why the hospital board is responsible for the quality of care provided by the hospital. These reasons are:

1. The ethical and moral obligation of the board;
2. The legal accountability of the board;
3. The accountability of the board to legislative and regulatory requirements; and
4. The fiduciary responsibility of the board.

The ethical and moral obligation of the board to assure quality of care is perhaps the most important but least recognized board mandate for quality. Because the vast majority of boards represent the community to the hospital and the hospital to the community, it is the trustees' family, neighbors, friends, business associates, or customers who are patients in and served by the hospital. From this and the fact that the board is the ultimate authority in the hospital flows the board's ethical and moral obligation to assure that quality care is provided.

The legal accountability of the board to assure the provision of quality care and to oversee the activities of the medical staff was reviewed in detail in chapter 3. Boards should be educated about such legal decisions as *Bing v. Thunig, Darling v. Charleston Community Memorial Hospital, Johnson v. Misericordia Community Hospital,* and others relevant to their specific states. Trustees must understand the consequences of inadequate or nonexistent board oversight of quality and the medical staff, both in terms of legal liability and severe and preventable patient injuries. The other side of this legal liability, antitrust lawsuits filed against the hospital by physicians as exemplified by the case of *Patrick v. Burget,* should also be used to educate the board about the danger of rubber-stamping medical staff admissions or credentialing recommendations.

Legislative and regulatory requirements of the board were also reviewed in detail in chapter 3. The board should be made aware of the language of their specific state hospital licensure laws and department of health requirements that specify the board's authority and responsibility regarding quality and medical staff oversight. Further, Joint Commission Standards and Conditions of Participation in Medicare requirements relating to board authority and responsibility should also be reviewed with boards.

The final reason a board is responsible for quality stems from their fiduciary duty to the hospital to act with reasonable diligence, care, and skill to protect and preserve the hospital's assets and financial viability. As many trustees regard their primary responsibility as being the financial well being of the hospital, it is important for the board to know that poor quality can have significant negative financial impact for the hospital. Poor quality care can facilitate malpractice liability as well as negative community perception of the hospital which can consequently lower admissions and revenues. Thus, boards should realize that effective quality and medical staff oversight can help preserve and advance the financial reserves and future of the institution, which is especially important in the current competitive health care environment.

What Is Quality?

Once a board understands why it is responsible for quality it should address the issue of what quality is. Defining quality is a difficult exercise and is best approached by examining and discussing the many perspectives of quality. There is, for example, the medical perspective of quality which is employed by physicians and may relate to such variables as the degree to which care meets current accepted practices measured by clinical research and monitoring and evaluation. There are the perspectives of the patients and their families regarding quality which may relate to how they were treated personally by hospital staff, to waiting times and amenities and such measured by patient complaints and patient surveys. There are many perspectives of quality including: statistical/demographic, technical, third party payor, government, Joint Commission, malpractice attorneys and case law, media, community, hospital managers, hospital staff and employees.

The reason that the various perspectives of quality are important to a board is that the combination of perspectives used by the board to conceptualize quality forms for the board, and thus the hospital—the de facto definition of quality. That is, if the board only receives information about medical monitoring and evaluation, then the board will define quality only in medical terms. If the board's information about quality only relates to compliance with JCAHO standards, then the board will define quality only as meeting JCAHO standards and achieving and maintaining accreditation.

Quality is clearly difficult to define and measure. The board must understand, however, that no single perspective of quality is the ultimate and exclusive definition of quality. For example, the care provided to a group of emergency room patients may be of high quality from a medical perspective but of low quality from the patients' perspective. In this situation, the overall quality of care provided by the ER has elements of both good *and* bad quality. It would be a mistake to define it solely by either element or perspective.

Once a board recognizes that the perspective-specific information it receives in fact defines *its* overall perspective of quality, it can then begin to discuss what different perspectives of quality-related information it wants to receive on a regular basis. In deciding on what types of quality-related information to receive, the board and QA professional develop a hospital-specific definition of quality that is more meaningful than any externally-developed definition of quality. Generally, the more perspectives that the board incorporates into its definition of quality, the more meaningful that definition of quality will be. A broad-perspective approach to quality will necessitate many streams of quality-related information flowing to the board which, when properly presented by the QA professional, helps the board to more effectively integrate information to monitor and improve overall quality and to minimize the risk of patient injury.

How Quality Is Measured

Boards need to know the basics of how quality is measured in order to track the effectiveness of the QA program and to establish targets for quality improvement. This does not mean giving the board a presentation on severity-adjusted quality measurement systems and sending them detailed system printouts, even if such a system is operational in the hospital. Boards first need a grounding in the basics in order to get a firm grip and comfort level with the exclusive concept of quality.

Consequently, the board should be educated about the three basic components of quality measurement: structure, process, and outcome. Further, the board should be educated about past efforts to measure quality by focusing exclusively on structure and process, and the shortcomings of those efforts.

Structure relates to the integrity of the facility, the condition of the equipment, and the quality of the supplies. Process relates to *how* care is delivered and includes such variables as staffing patterns, implementation of policies and procedures, and medical technique. Outcome relates to the condition of the patient following some type of hospital intervention such as a diagnostic or a therapeutic procedure. It is important to stress that outcome does not simply refer to the patient's condition at discharge as a patient can conceivably have a distinct outcome for each intervention. In fact, a patient can have several intra-hospital stay outcomes, one outcome at discharge, and another related outcome several months post-discharge.

The fallacy of structure or process indicating outcome should be discussed with the board as a way to emphasize the importance of outcome-focused QA activities. It is possible to have good structure coupled with good process and yet have bad outcome. Conversely, it is also possible to have bad structure and bad process and still have good outcome. To be able to begin to make meaningful quality judgements, boards must firmly understand structure, process, and outcome and must receive clear measures of patient outcomes in their quality reports.

Once a board understands the importance of outcome measurement in QA it gains confidence in evaluating the quality of care and can begin to develop a mechanism to evaluate the effectiveness of the QA program as well. For example, in conjunction with the QA professional, a board might determine that 70 percent of the patient care measurement activities of the QA program should focus on patient outcomes. Thereafter, the 70 percent outcome focus becomes one target by which the board can judge the effectiveness of the QA program.

In addition to the three broad components of quality, it is important for a board to be familiar with the three time frames of quality measurement: retrospective, concurrent, and prospective. An understanding of the time frames of quality measurement, along with the relative strengths and weaknesses of each, will strengthen the board's grasp and comfort with the hospital QA program and the concept of quality and its measurement.

Retrospective review simply refers to measuring care that was provided in the past, generally after the patients have been discharged from the hospital. Retrospective QA attempts to assess the quality of care provided in the past, say, six months, to identify problems with that care or elements that can be improved. Then, in the present, that information is used to correct the problems or improve the elements of care that will be provided from that point on. Retrospective QA is mirrored after experimental research design methodology and as such has the appeal of scientific validity. Retrospective QA is powerful and useful but has a major limitation: the feedback loop from the time a problem is identified to the time a problem is corrected can be too long. This means that while a component of past care, say broncoscopy, is under investigation, patients in the present may be receiving poor quality broncoscopies and sustaining injuries. Boards need to know the strengths and weaknesses of retrospective QA, and then perhaps establish a maximum target, for example 60 percent, for the use of retrospective review in the QA program. Reviewing with boards when retrospective review is most appropriate, when complicated multi-variate issues of care are being investigated or when large numbers of patients are involved over long periods of time, will aid their understanding of both the power and limitation of this quality assessment approach.

The limitations of retrospective review can be overcome by the use of concurrent review. Concurrent review is defined as the review or assessment of care *as* it is being delivered. As this is clearly not always possible, a good working definition of concurrent review is the assessment of care while the patient is still in the hospital, say within 24 (or 48, or perhaps 72) hours of its being delivered (Longo et al, 1989). The obvious value of concurrent review is the short feedback loop from the time a problem with patient care is identified to the time it is corrected. Boards should be given examples of concurrent review mechanisms currently in use in the hospital such as incident or occurrence reporting, use of quality indicators, and certain generic screening systems to fully understand the power and applications of concurrent review. Once a board grasps both retrospective and concurrent review, it will support the QA professional's efforts to move the focus of the QA program away from retrospective review and toward concurrent review.

Prospective review is a bit more difficult for a board (and many QA professionals) to conceptualize. Prospective means in the future, review means in the past, so how does one do prospective review? Prospective QA simply means the review of intended care; the examination of new procedures, techniques, or systems *before* they are implemented to make certain as many potential problems as possible are anticipated and *prevented* from occurring. Prospective QA does occur as a routine event at most medical and nursing departments but is rarely recognized as QA or incorporated into the QA program.

Prospective QA is, however, a very powerful QA approach. Since it focuses on preventing problems and making certain the quality of care is as good as it can be, boards should be made very comfortable with the concept. In fact, unlike any other QA activity, prospective QA is an activity that should be routinely performed *by* the hospital board.

Hospital boards regularly make decisions about new business activities or the purchase of new equipment. These decisions may involve establishing hospital-sponsored home care, setting up a managed care program, purchasing a laser surgery instrument, or an MRI unit. When making this type of decision, the typical board considers the financial implications, the capital requirements, a marketing plan, return on investment, and many other variables associated with the new venture. Unfortunately, rarely does a board consider how the new activity will be integrated into the hospital's QA program or its potential impact on the quality of care. By insisting that a quality plan be a part of all such major decisions a board actually *performs* prospective QA. By doing so, the board sends a message throughout the hospital that quality is a top priority and helps to ensure that potential quality problems will be identified and prevented *before* they can have a negative impact on patient care. Thus, the prospective approach to quality is a powerful and critically important one, and one that is uniquely in the board's domain (Orlikoff, 1989).

What Is Quality Assurance?

Once a board has a grounding in the concepts of quality and its measurement, the next step is to review the general definition and purpose of QA and the specifics of the hospital's QA program.

Quality assurance can be defined as the examination and evaluation of patient care, and factors that affect patient care, toward the goal of continuingly improving the quality of that care. Thus, QA is composed of two general components:

1. The problem-focused approach—activities designed to eliminate or reduce problems in patient care or bad outcomes.

2. The continuous improvement approach—activities designed to make good (i.e. problem-free) quality patient care even better.

These two components are critical to both board understanding of QA as well as to effective board oversight of the QA program. For example, discussing and determining what percentage of the QA program should be devoted to the problem-focused approach and what percentage to the continuous improvement approach is a useful exercise for the board to grasp QA. It is also very beneficial to the effectiveness of the board's oversight of the QA program by, again, establishing targets for the board to assess the QA program and helping the QA program retain a sharp focus.

This discussion should then lead into the specifics of the hospital's QA program which should include: the value of the QA program; its hospital-wide scope; the various components of the program; the organization of the program; medical staff involvement in the QA program; nursing and all departments and services involvement; and, briefly, related activities to QA such as infection control, utilization review, safety, and risk management (JCAHO, 1988).

The hospital QA program is the flagship in the hospital's effort to deliver the quality of care it wants, as determined by its governing board, to provide. Through activities designed to both identify and resolve problems in patient care and to continuously improve problem-free care, the QA program should function throughout every area in the hospital. The efforts necessary to effectively integrate QA activities throughout the hospital should now be clear to the board; as should the necessity for board commitment to quality, and board support for and oversight of the QA program.

Why Do Quality Assurance?

When a board understands the basics of quality assurance it will realize that quality can indeed be defined, measured, and improved. At this point, assuming that the board has been reasonably educated about quality and QA, they should understand quite a bit about quality and, more significantly, be able to begin effectively overseeing the QA program. For example, the board, perhaps with the help of the QA professional, has at this point framed a definition of quality for itself and the hospital. This should be reflected by the types of information the board now plans to receive regarding quality. The board and QA professional may have determined that the hospital's quality measurement focus needs to shift more to outcome measurement, and may have set a specific target figure, say, that 70 percent of all quality measurement activities should be outcome-focused. Further, the board may desire the quality measurement activities to become more oriented to concurrent review as opposed to retrospective review. And the board may itself, for the first time, routinely engage in prospective QA by integrating quality into all of its major decisions. If this occurs, the board is well on its way to developing a sincere commitment to quality and to QA, unless the board is unclear or misinformed about the purpose of quality assurance.

Unfortunately many hospital CEOs, medical staff members, and even QA professionals believe that the primary purpose of QA is to gain accreditation by the Joint Commission on Accreditation of Healthcare Organizations or to satisfy other regulatory or legal requirements. This attitude is counterproductive to the true purpose of QA and creates a defensive approach to quality in the institution. Defensive quality is a very limiting concept, as it motivates those in the hospital to do only enough to satisfy regulatory and legal requirements. Once those requirements are satisfied, the QA efforts slack off as the "goal" has been attained. Paradoxically, the attitude of defensive quality actually serves to elicit greater, more stringent regulation. As the regulators see hospital QA programs only doing enough to get by and not really attempting to improve quality, they develop more numerous and stringent requirements.

The hospital board must rise above the dangerously limiting perspective of defensive quality. It is incumbent upon the QA professional to first make certain that *they* are not afflicted with this misconception, and then to make certain the board does not succumb to it.

The pure and simple purpose of quality assurance is to improve the quality of patient care. The board must discuss this concept, come to truly believe it, and work to instill this belief throughout the hospital if there is to be a meaningful commitment to quality and its continuous improvement. The QA professional must be a champion of this perspective in order to positively influence the board and to gain sincere board support for their activities.

The Problem-Focused QA Process

Although much has been written recently about moving QA away from reliance on the problem-focused approach, it still has great value in the QA armamentarium. Furthermore, the problem-focused QA process, and where it tends to break down, is very useful for boards to know. It gives boards both a method to evaluate QA program effectiveness as well as an understandable format for the presentation of QA information.

The problem-focused QA approach is best expressed as a five-step process. There must be steps to:

1. Identify problems
2. Assess problems—determine the cause and scope of the problem
3. Implement action to correct the problem
4. Monitor the corrected problem to assure that the desired result has been achieved and sustained
5. Document the process and follow up as appropriate.

This process is useful to boards because any quality problem, regardless of its complexity or clinical detail, can be analyzed by and tracked according to its progress through the steps of the process. It is important to note that, like QA itself, the board does not actually perform the steps of the problem-focused process. Rather, the process is useful to the board as a framework from which to regularly review information related to quality problems in the hospital.

This type of review indicates to the board how quality is being improved via the elimination and reduction of problems, as well as how effectively the QA program is applying the process throughout the hospital. For example, a board receiving summary QA problem-focused information presented in the format of the problem-focused process (i.e., how many hospital-wide problems were identified last quarter, how many were assessed, and so on) might discover that far more problems have been identified than have been assessed or corrected. Or, a board might discern a pattern where "corrected" problems are not monitored to assure that the problems have in fact been permanently eliminated or sufficiently ameliorated. In both of these situations the board is able to direct resources or organizational authority to ensure that all steps of the process are completed whenever possible. Thus the board is able to effectively and appropriately discharge its oversight role and have a positive impact on quality. Moreover, the board is also able to direct resources and support to the QA professional and QA program when it perceives that such resources are necessary to upgrade QA efforts to target levels.

THE MONITORING AND EVALUATION PROCESS

Although less useful than the problem-focused process in terms of a format for presenting QA information to the board, the monitoring and evaluation process is regarded as "the cornerstone of effective quality assurance activities" (JCAHO, 1988). As such it is useful to present this process to the board as background so that the board can periodically verify that the complete process is being employed continually throughout all medical staff department QA activities, all hospital-wide QA activities, and clinical and support service QA activities.

The JCAHO's ten-step monitoring and evaluation process is as follows (JCAHO, 1988):

1. Assign responsibility for monitoring and evaluation activities

2. Delinate scope of care

3. Identify important aspects of care (high-risk, high-volume or aspects prone to problems)

4. Identify indicators related to those aspects of care

5. Establish thresholds for evaluation related to the indicators

6. Collect and organize data

7. Evaluate care when thresholds are crossed

8. Take actions to solve identified problems

9. Assess the actions and document improvement

10. Communicate relevant information to the organizationwide quality assurance program.

It should be pointed out to boards that the ten-step monitoring and evaluation process subsumes the five-step problem-focused process presented earlier. While boards should be presented with the monitoring and evaluation process, they should not be expected to routinely monitor to verify that the process is in place and working effectively. This can be done periodically, such as every six months. The board's more frequent monitoring of the problem-focused process, however, will also indicate the effectiveness of the functioning of the monitoring and evaluation process. Furthermore, the board will find the problem-focused process easier to monitor and understand, and thus will be more likely to deserve regular attention to it than to the monitoring and evaluation process.

The quality indicators aspect of the monitoring and evaluation process (step #4) and the thresholds (step #5) should, however, be reviewed in some detail with the board. Equating the indicators and thresholds to the problem identification step of the problem-focused process will be quite useful to boards as they gain an understanding of how specific aspects of structure, process, and outcome can be measured. In fact, the board may wish to periodically monitor patterns of certain significant hospital quality indicators to ascertain that problems are in fact being identified and solved and that opportunities to improve the quality of care are being exploited.

To summarize, the monitoring and evaluation process should be presented to the board as QA background, but should not be used as a format for organizing and presenting routine QA information to the board. Nor should the board be encouraged to routinely oversee the monitoring and evaluation process. This is because the monitoring and evaluation process, while a key QA activity, is too detail-specific and complicated to be routinely useful to the board as *governance* QA information. Rather, the problem-focused process, which is actually a distillation of the key points of the monitoring and evaluation process, should be used as one format for the routine presentation of QA information to the board.

Distinguishing QA Program Information from Information about Quality

It should be made clear to the board that it must review two broad categories of information: one that relates to assessing and improving the quality of care as such, and one that relates to assessing and improving the effectiveness of the hospital-wide QA program. Although these two types of information will obviously interrelate and occasionally be difficult to distinguish from one another, it is important that the board understands that it must perform these two distinct activities and that it knows which activity is which relative to each quality-related report it receives.

Some information will only facilitate assessment of the quality of care and some will only facilitate judgements about the effectiveness of the QA program. Often, however, quality-related information will contain elements indicating *both* quality and the functioning of the QA program. In these frequent situations it is incumbent upon the QA professional to make certain the board is aware of the dual nature of the information and to identify when circumstances in one set of information may mask or lead to erroneous conclusions about the other.

This situation is often seen, for example, when a QA program's problem identification activities are reported to the board. If the problem identification techniques are ineffective or marginally functional few patterns of significant problems will be identified and reported to the board. Here, the board may incorrectly conclude that there are no patterns of problems in patient care, that the quality of care is at acceptable levels, *and* that the problem identification component of the QA program is functioning effectively.

Conversely, in the situation where the problem identification techniques or their application have been improved, more problems will be identified and the board may incorrectly conclude that the quality of care has deteriorated. In fact, the correct analysis is simply that the QA program component of identifying problems has improved. As accurate information on patterns of problems is being generated for the first time, no conclusions about quality trends can be made until more data is gathered over time by the improved problem identification techniques.

It is important that the board be made aware of the differences between QA program information and information about quality of care. It is critically important, however, that the QA professional be aware of these differences and constantly monitor for situations where the two types of information mingle together and influence clear analysis of or mask one another. The QA professional must then identify and clarify these situations for the board and provide guidance in their analysis.

QUALITY ASSURANCE PROGRAM INFORMATION AND EFFECTIVE BOARD OVERSIGHT

When the board is familiar with its approach to quality, and the purposes and several of the general techniques and processes of the QA program, it is time to review key information about specifics of the QA program itself with the board. This review should provide the board with a broad perspective of the structure and function of the hospital's QA program. Then, the board should discuss and come to an agreement as to how it will exercise oversight of the QA program.

Effective board oversight of the QA program is composed of two elements: meaningful governance information to the board about the functioning of the QA program; and, appropriate board action in response to that information.

One of the key informational items and guideline documents about the QA program is the QA plan. A viable QA plan will facilitate the consistent application of effective QA processes throughout the hospital by specifying the organization, function, and activities of the QA program as well as the roles of all groups involved. Unfortunately, most boards are strikingly unfamiliar with their hospital's QA plan—even though they probably have voted to approve it. The QA plan should be introduced to the board as a working document that should be revised often to effect or reflect improvements in the structure and function of the QA program. The board should be educated to recognize that the purpose of the QA plan is not simply or exclusively to comply with JCAHO requirements.

A good QA plan will be a very useful reference document for a hospital board as it should specify the relative roles of the governing board, medical staff, management and QA professionals in the QA program. This means that once a board has defined or refined its role that it should be incorporated into the QA plan. Further, if a board needs clarification about its role in QA, a good QA plan will provide answers and should be consulted.

Equally important, the QA plan should contain sufficient detail to support board assessment of the QA program. By stating what QA activities are to be performed by whom and how often, and how and when QA information should be reported, a good QA plan will facilitate effective board oversight and evaluation of the QA program. Consequently, each board member should be provided a copy of the hospital's QA plan for background, and copies of the plan should be available for consultation during board meetings when quality issues or the QA program is being discussed.

The board should also be involved in the annual appraisal of the QA program and the QA plan should both be used to help make the appraisal as well as be revised to reflect resulting modifications in the QA program. The board should be presented with the goals of the QA program reappraisal and with the methods for conducting, and components of, the reappraisal. The Joint Commission states that the annual reappraisal of the QA program should include the following (JCAHO, 1988):

- assessment of the monitoring and evaluation process to determine its effectiveness;
- comparison of the written (QA) plan with the quality assurance activities that were performed;
- determination of whether quality assurance information was communicated accurately and to the appropriate persons, committees, or other groups; and
- determination of whether identified problems were resolved and patient care improved.

Participation of the board in the annual appraisal of the QA program is an excellent way to achieve effective board oversight without inappropriate board over-involvement. The board is able to determine specific QA program strengths and weaknesses and assess and improve the overall QA program by stepping back and comparing the program to the QA plan, the information flow, and whether problems have been consistently identified and resolved and patient care improved over the past year. This is an excellent oversight activity and is an especially appropriate and rewarding one for the board.

As the board evaluates the QA program during the annual appraisal, and throughout the year, it should do so to verify that four critical objectives are achieved. First, the QA program must be hospital-wide. This means that all departments, areas, and services, and all individuals whether employees or independent contractors, that provide care to or interact with patients must participate in organized QA activities. All individuals who provide care must have that care and their performance evaluated. The importance of this must be impressed upon the board and the point can easily be made by citing several egregious medical malpractice cases that resulted when health care providers' activities were not evaluated by routine QA functions.

Second, the board should review the QA program to verify that it focuses on identifying and resolving problems. As discussed earlier, the board may have established a target percentage of all QA activities that should be problem-focused. The board should verify that the target was met, and determine annually if the target level is appropriate or should be raised, or, more infrequently, lowered.

Third, the board should review the QA program to assure that it is integrated with other quality-related activities of the hospital. These other quality-related activities include but are not limited to: medical staff appointments and privilege delineations, risk management, infection control, patient representative programs, and safety and security programs. The board should be provided with information that verifies: appropriate information flow between these activities; that there is a central coordinating function that includes and integrates all such activities; that there is no duplication of effort, such as the same problem being investigated by two different quality-related activities without knowledge of each other; and that the different activities can integrate to work together to solve problems and improve care when appropriate.

Fourth, and by far the most important, the board should review information about the QA program that demonstrates that it has *improved* the quality of patient care.

This is the primary and most important purpose of the QA program and so it should be the primary and most important purpose of the board's review of the QA program.

The remainder of this chapter will review other, more specific QA and quality-related data that comes primarily from internal sources.

Clinical Data

During its orientation to or review of the QA program the board should be exposed to the many different types of routine departmental and medical staff monitoring and evaluation activities and functions. These activities include: departmental review, morbidity and mortality review, surgical case review, blood utilization review, drug utilization review, medical records review, and pharmacy and therapeutics review, and other review functions specific to individual hospitals.

Usually, the majority of a board's members are quite surprised to find so many reviews continuously occurring within the hospital. The intricacies of QA and the considerable support and expense necessary to maintain these functions effectively should now be clear to the board. The board should now also realize the incredible wealth of data the hospital QA professionals and medical staff have access to, if properly supported. The board must also be educated to realize that it means nothing that every single surgery in the hospital, or every death, is reviewed unless the reviews are effectively and honestly conducted to generate meaningful data, and unless the data is then put to use to correct problems and improve the quality of clinical care.

This is a key point and must be impressed upon boards. The goal is not simply to conduct required reviews, nor to generate paper, nor to give a cursory nod that the care is great and everyone is doing a fine job. The purpose of the reviews is to screen for problems and to see if those problems are significant and if they cluster in identifiable trends and then to initiate action to correct the problems. These reviews should be explained to the board as being examples of the hospital's problem identification and assessment techniques. They should then be evaluated by the board to determine if in fact the reviews are functioning effectively to identify and assess problems or if they are simply generating piles of wasted paper.

The QA professional must work with the board, and medical staff, to reach agreement on how to distinguish effective clinical reviews from ineffective ones that actually obfuscate problems rather than identify and solve them. Symptoms of this are, for example, one years worth of surgical case reviews where all or most variations from the screening criteria are later justified as being acceptable. In this situation, either the screening criteria are too rigid, or the review process for variations from the criteria is too loose and not being scrutinized appropriately.

Once the board verifies that the clinical reviews are functioning effectively, or have been corrected, they should then receive periodic *summary* information from selected review functions. Those reviews reflecting complicated or highly invasive care or situations of potential patient injury, such as surgical case review and morbidity and mortality review, may warrant more frequent reporting to the board than more administrative reviews such as medical records reviews. To facilitate the board's analysis of such information, the QA professional should develop reports that present trends over time so that the board can track progress toward preestablished goals, and compare the quality of care against care provided by the hospital in the past. Report formats for QA information to the board, and their critical importance, will be discussed in greater detail in chapter 6.

Granting Medical Staff Appointments and Privileges

Perhaps the single most important QA responsibility of the hospital governing board is ensuring the appointment and appropriate retention of a qualified medical staff. Because the overall quality of a hospital's care is extraordinarily influenced by the composition and performance of its medical staff, and because the board is ultimately accountable for both the performance of the medical staff and the quality of care provided, boards must be convinced to abandon the traditional, and unfortunately still prevalent, "rubber-stamp" approach to granting medical staff appointments and delineating clinical privileges. The rubber-stamp approach refers to boards that simply vote to approve, or "rubber-stamp," whatever the Medical Staff Executive Committee recommends without any discussion or attempt to verify that the medical staff employed a meaningful, objective, criteria-based process to evaluate candidates.

The credentialing decisions made by a board will have a significant impact on both the present and future quality of care provided at its hospital. Consequently, QA professionals must work with their boards to educate them about the critical importance of this activity and to make them more comfortable performing it by giving them an understandable purpose for the activity and a clear process to follow.

Boards that rubber-stamp medical staff credentialing decisions tend to do so for two basic reasons: they are intimidated by the medical staff or are afraid of generating or exacerbating conflict with the medical staff; and, they have absolutely no notion of *how* they should be reviewing the recommendations of the medical staff and so employ no organized process for doing so.

The concern of generating conflict with the medical staff was discussed in chapter 3 and is best addressed by fully educating both board and medical staff about their respective roles, and by developing an organized process for the board to use in considering and voting on medical staff recommendations for admission to the medical staff and privilege delineations. Further, the board must completely grasp *why* it is ultimately responsible for medical staff credentialing, also covered in chapter 3, and become comfortable with that intimidating but crucial responsibility.

To accomplish this, the board must first understand that appointments (both initial appointment and reappointment) to the medical staff and the delineation of clinical privileges are two separate processes. The purpose of the appointment process, both initial and reappointment, is to determine the applicant's eligibility to be a member of the hospital's medical staff. The information necessary to make this decision relates to education and training, licensure, and experience. What is being determined is the *capability* of the applicant to practice medicine. When clinical privileges are granted, however, the issue is insuring that the applicant is *competent* to practice at the skill level requested and the information required relates to professional performance, specialized training, and references (Orlikoff and Vanagunas, 1988).

In both processes the board, medical staff, and management work together to reach the best decisions regarding which physicians should be admitted or readmitted to the medical staff *and* what procedures and activities they will be allowed to practice and conduct. Although it is a collaborative process, the board must verify that the medical staff and management have performed their roles accurately and completely.

The relative roles of medical staff, management and QA professionals, and the board regarding medical staff admission, readmission, and privilege delineation are as follows (Totten and Orlikoff, 1987):

I. *Role of the Medical Staff*
   1. Explains evaluation process to applicant.
   2. Assesses each applicant objectively and consistently against written criteria for:
      - professional competence
      - acceptable clinical performance
      - character and fitness

3. Recommends each applicant for approval or denial to the hospital governing board.

II. *Role of Administration—QA Professionals*
1. Provide adequate resources and support for the medical staff's evaluation process.
2. Facilities information flow from applicant to medical staff and back, and from medical staff to board and back.
3. Often assists medical staff in information collection and verification of applicant's credentials, licensure, insurance, past hospital affiliations, references, and the like.

III. *Role of the Governing Board*
1. Assures that the medical staff bylaws contain a specific, criteria-based procedure to evaluate applicants.
2. Verifies that the procedure is followed by the medical staff equally, completely, consistently, and objectively for each applicant's evaluation.
3. Confirms that the appropriate information for each applicant has been reviewed and verified.
4. Reviews each applicant thoroughly by comparing the medical staff's recommendation against the criteria for medical staff membership and privilege delineation.
5. Based on its review, makes final decision on each applicant.

Once the relative roles and final authority are clear to the board, it is time to make the important legal and political distinctions between *initial* appointments to the medical staff and privilege delineations, and *reappointments* to the medical staff and renewal of privileges.

In initial appointments the process relies on external data sources such as the hospitals at which the applicant previously practiced. The burden of proof of competence in this process is *on the applicant* and this should be stressed to the board as it is much easier to keep a questionable or incompetent physician off the medical staff than it is to remove him from it once he or she is admitted to it.

Initial appointment confers staff membership to the physician and, via privilege delineation, develops what is in essence a job description for the physician. Initial appointment also triggers a key and often underused period of provisional membership where the physician's performance is supposed to be closely monitored by his department and the medical staff credentials committee. This is the best time to catch potential problem physicians and remove them from the medical staff before they are protected by the complete due process procedures afforded by full medical staff membership. Or, their privileges can be limited before full membership is approved by the board. Boards must insist that the medical staff put the provisional period to good use and diligently monitor and honestly report back on the performance of physicians with provisional status.

Reappointment of physicians to the medical staff and renewal of their privileges, on the other hand, relies on internal data sources. The QA monitoring and evaluation activities that have tracked and evaluated the physician's performance at the hospital *must* be funneled into the reappointment deliberations of the medical staff and be verified by the board. This is where major breakdowns in the system typically occur; when clinical QA data that yields information on the quality of a physician's care and on his or her skill is *not* integrated into the reappointment process—the process becomes a meaningless paper exercise. The reappointment process is supposed to emphasize the evaluation of the performance of the physician at the hospital, and to do this the clinical QA data must be used. It is absolutely critical that the board understand and insist upon this.

This performance-based review is crucial because during the reappointment process the burden of proof is on the hospital to clearly demonstrate why a physician should be denied readmission or have privileges restricted. Hence, it is easier to *prevent* the initial admission of an unqualified or badly skilled physician to the medical staff than to *remove* one from the staff. For this reason the critical nature of both the initial appointment process and the need for performance-based evaluation during the reappointment process must be clearly understood by the board. The board should be informed that the readmission process must be conducted at least every two years for each medical staff member and that the process is essential to comply with Joint Commission and other regulatory standards.

To facilitate reaching appropriate final decisions for both initial and reappointments the board should consider a manageable number of cases at one time and schedule them at regular intervals, perhaps as often as every meeting depending on the size of the medical staff. This works much better than the board that considers thirty or more applications or reapplications at one time; in this situation the temptation to make rubber-stamp decisions is almost irresistible to any board no matter how motivated it may be. Furthermore, the board must ensure that all applications are complete and that all questions regarding an applicant have been answered satisfactorily. And the board must, with the assistance of the QA professional, develop and adhere to a mechanism for making final decisions.

An effective mechanism for making decisions is the key to achieving both trustee comfort with their oversight responsibility for the medical staff credentialing process and to their meaningfully performing the function. The mechanism must reflect the fundamental distinction between the medical staff's role and the board's role which is simply this: the medical staff evaluates each applicant; the board monitors to assure that the medical staff has performed each evaluation properly and completely. This, when distilled to its essence, is a very comforting and acceptable division of responsibility for both physicians and trustees.

The trustees do not evaluate the skill, training, ability,

or performance of the physicians—the medical staff does that. The board simply evaluates the *process* used by the medical staff to evaluate applicants and only intervenes if the process has been breached or incompletely or inconsistently applied. In order to evaluate the process, however, the board must know *how* the medical staff evaluates the applicants; the board must know what criteria are used.

This raises the concept of decision criteria. Those measures that are used by the medical staff to evaluate each applicant should also be used by the board to monitor the evaluation process. These decision criteria should be developed jointly between board and medical staff with the medical staff taking the lead. The criteria are then incorporated into the bylaws of the hospital medical staff and must be consistently and completely applied by the medical staff department heads, credentials, and medical executive committees to all applicants. This helps make the evaluation process used by the medical staff, and its recommendations to the board, more objective.

The board then must confirm that the criteria were completely and consistently applied to all applicants, and verify that the recommendation of the medical staff for each applicant was consistent with the criteria and therefore the correct recommendation.

When the board verifies that the recommendation of the medical staff executive committee is consistent with the criteria, the board votes to approve the recommendation regarding the applicant. If the recommendation of the medical executive committee is inconsistent with the criteria, however, then the board reverses the recommendation and takes a vote that is consistent with the criteria.

In this manner, the board is able to assess objectively the evaluation process used by the medical staff and to assure, via correct use of its final decision authority, that the proper decisions regarding medical staff admissions and privilege delineations are consistently made. The key to this process, and to board comfort with it, is that the board does not really make a decision as such; the *criteria* make the decision for the board—as they should for the medical staff.

If, however, the medical staff makes recommendations that are inconsistent with the criteria the board must take two actions. First, the board votes on each applicant in a manner consistent with the criteria and reverses the medical staff recommendation.

Second, the board then communicates with the medical staff departments, credentials and executive committee to inform them that they are not performing the applicant evaluation process properly and to instruct them to do so by adhering to the criteria. Thus the board both effectively oversees the medical staff's applicant evaluation and privilege delineation process as well assures that only qualified, skilled, and competent physicians are admitted to the medical staff and that their privilege delineations are appropriate.

Two essential keys to this process are clear and objective decision criteria, and a concise mechanism for communicating the medical staff's recommendation for each applicant along with a summary sheet that details the applicant's compliance with the decision criteria to the board.

Regarding clear and objective decision criteria, it is useful for both the medical staff and the board to employ two categories of criteria: Level 1 and Level 2 criteria. Level 1 criteria are "dealbreaker" criteria, meaning that each Level 1 criteria *must* be met by an applicant in order to be admitted or readmitted to the medical staff or to have privileges granted. Examples of Level 1 criteria are: valid state medical license; verification of medical school education, residency, and other required training; current and adequate malpractice insurance; and, a completed application.

Level 2 criteria are those criteria where failing to meet any single criterion is insufficient reason to deny admission or readmission to the medical staff or deny or restrict privileges. However, an applicant's failure to comply with, say, three of the ten Level 2 criteria would be the threshold for denying admission or restricting privileges. Examples of

Level 2 criteria are: adequate use of privileges previously granted; no (or specified maximum number of) past or pending disciplinary actions by hospital, professional societies, or specialty boards; specified negative patterns of adverse clinical outcomes; and, no reductions or revocations of privileges at other hospitals.

Clearly, any Level 2 criterion can be elevated to a Level 1 status. The decision is up to the medical staff and board. What is important is that the criteria are made explicit and objective and then consistently used. The advantage of using the Level 1 and Level 2 criteria model is that it makes the process more objective and important criteria that are in and of themselves insufficient to base an admission decision on can still be used in the evaluation process as Level 2 criteria.

Regarding communicating the information to the board for its vote, the board should receive a summary sheet for each applicant. This sheet should contain the recommendation of the appropriate medical staff committees(s), and a listing of all Level 1 and Level 2 criteria by category and verification that each criterion was or was not met. Thus, the board can quickly compare the medical staff's recommendation to the applicant's compliance with the criteria and render an appropriate decision. The issue of report formats to the board will be addressed further in chapter 6.

To assure that medical staff credentialing is effective and results in a fair and rigorous examination of each applicant, the board must understand the process and its role in it. The board must actively participate in the process by assuring that all evaluations have been properly conducted and that its final decision is criteria-based and therefore as objective as possible.

INFECTION CONTROL

Boards should be made aware of the concept of nosocomial infections, those infections that are acquired by or afflicted upon the patient in the hospital. The seriousness of nosocomial infections both in terms of frequency and severity should be impressed upon boards to emphasize the importance of the hospital's infection control function.

Although various studies have indicated a wide range in the incidence of nosocomial infection, a reliable *minimum* figure is that 5 to 6 percent of all patients in acute-care general hospitals sustain nosocomial infections. This minimum infection rate (some studies place the figure as high as 20 percent) generates infections in somewhere between 1.5 to 2.1 million patients annually. These infections result in or contribute to between *20,000 to 80,000* patient deaths annually, which places nosocomial infections in the top ten causes of death in the United States (JCAHO, 1988; Haley et al, 1987; Haley, 1986; Holzman, 1988; Missouri Professional Liability Insurance Association, 1978).

Boards should be made aware of both the seriousness of nosocomial infections as well as their staggering financial implications. Holzman (1988) states that the typical nosocomial infection that does not result in death increases the patient's stay in the hospital by four days and costs the hospital approximately $1,800. Various sources estimate the annual aggregate cost of nosocomial infections to be between 2 to 4 billion dollars (Haley et al, 1987; Missouri Professional Liability Insurance Association, 1978).

Even more disconcerting is the probability that the incidence of nosocomial infections is likely to increase in the future (Longo et al, 1988). This is due to many factors, some within and some beyond the control of the hospital, including: nursing and staff shortages; increased use of invasive procedures; more antibiotic-resistant organisms; more immunologically suppressed patients—such as those with AIDS; and the increasing elderly population.

All of these factors should convince the board of the importance of the infection control program, the value in integrating it into QA and medical staff credentialing, and the need for board oversight of the activity.

Initial reports to the board should frame the nature of

the problem in the hospital by identifying the most frequently occurring types of infections, areas of occurrence, and implications in terms of patient outcomes and cost to the hospital. Then the activities of the Infection Control Committee should be reviewed by the board to verify that the hospital-wide level of infection surveillance and control is appropriate and effective. The past and current hospital-wide nosocomial infection rate should be identified for the board, target rates established, and the board should then receive periodic graphic information showing trends in infection rates relative to the target rate.

By monitoring and oversight, the board can verify that the infection control program is functioning effectively to *prevent and reduce* nosocomial infections, is operating concurrently, and is integrated in the overall QA program. Continued emphasis and commitment of board attention and hospital resources is necessary to maintain this key QA activity at acceptable levels of operation and results.

RISK MANAGEMENT

As the ultimate authority of the hospital the board is ultimately responsible for risk management. Yet, the only contact many boards have with risk management is when they vote to approve the purchase of insurance for the hospital or when they are informed of a large malpractice judgement.

Risk management relates to identifying and minimizing areas of potential patient injury that could generate malpractice losses for the hospital. More broadly, risk management also relates to risk financing, insurance, and claims management. The focus of risk management, however, on patient injury prevention and monitoring inexorably intertwines it with the activities of QA and medical staff credentialing. As such risk management (RM) should be integrated into the board's activities, and into board information reports on quality.

First, the board needs a brief introduction and clear description of the purposes and functions of risk management. This should include distinctions between its major activities such as loss prevention, claims management, insurance issues and so on. The organizational and conceptual relationship between QA and RM in the hospital should be reviewed with the board; and the board should make certain that the two activities are integrated as much as possible.

An extremely useful method for orienting trustees in RM, establishing the link between QA and RM, and for providing the board with a mechanism to track RM information and program effectiveness is to review the differences between iatrogenic and custodial patient injury.

Custodially-related patient injuries are those that do not specifically relate to or are caused by medical action or inaction. Rather, these types of patient (or visitor, or even staff) injuries are caused by custodial or administrative circumstance, action, or inaction. Typical types of custodially-related patient injuries include such events as: falls out of bed; slips and falls on wet or cluttered hallways; falls in the bathroom; and cuts or bruises suffered from broken or unsecured equipment or instruments.

Iatrogenic, or medically-related, patient injuries are those injuries that are directly related to or caused by physician or other professional health provider action or inaction. Typical examples of iatrogenic patient injury include: the incorrect body part or organ operated upon or removed during surgery; the incorrect patient operated upon; the perforation of a patient's bronchial tube during a bronchoscopy; a mis-read EKG which results in the patients discharge from the emergency room and a subsequent acute myocardial infection and death; an anaphylactic reaction to a drug or chemical agent; and nosocomial infections.

Custodially-related injuries are the most frequently occurring type of injuries in the hospital with somewhere between 75 to 85 percent of all patient injuries falling in this category. Custodially-related injuries tend, however, to be less severe than iatrogenic patient injuries. In terms of malpractice dollars lost, one measure of the severity of injuries, custodially-related injuries generally only account for approximately 15 to 25 percent of a hospital's malpractice losses.

These distinctions are important for a board to know in that they will help the board determine and direct the RM problem identification and resolution systems and integrate them into QA. For example, once the distinction is clear and the board has been informed about the RM's incident reporting system the board can determine if the incident or occurrence reporting system is focused appropriately. Many incident reporting systems focus on custodially-related patient injury almost to the exclusion of iatrogenic injury. When a board learns this it is generally surprised, recognizes it as inappropriate, and attempts to direct a refocusing of the system to iatrogenic patient injury.

Boards should also be made aware of the areas of greatest frequency of iatrogenic injury to patients within the hospital. Few boards are currently aware of this important quality-related information. This is critical information for a board as it identifies those departments and areas that require resources and oversight to improve the RM and QA activities and reduce the incidence of severe patient injuries, thus improving the quality of care.

In order for the board to oversee the RM program it must be informed of the activities and techniques of the program and receive regular reports on them. These can include pattern analyses of incident reports, summaries of malpractice claims, settlements, and lawsuits, and reviews of insurance coverage. The QA professional should make an effort to assure that the board understands the value in integrating QA and RM by demonstrating that many RM activities such as incident reporting can be viewed as QA problem identification and assessment techniques. Accordingly, QA and RM data should be integrated whenever possible in reports to the board.

Patient Satisfaction Information

Information that indicates how patients regard the hospital, its physicians and employees is also very useful for the board to periodically receive. This is important to reinforce to the board the many important perspectives of quality, as discussed earlier in this chapter, and will help the board maintain an integrated perspective on quality. Equally important, this type of information will keep the board in touch with the most important people in the hospital: the patients.

The board should be made aware of what type of patient-centered systems exist in the hospital, such as a patient representative program. Further, the board should also know what type of activities are conducted, such as patient complaint analyses or patient satisfaction studies.

The board should then receive reports such as those that analyze patterns of patient complaints or give results of patient satisfaction studies or report on the activities of the patient representative or similar program. These reports will enable the board to determine whether the care the hospital is providing is meeting the needs and quality expectations of those who receive it.

Hospital Staffing Patterns and Plans

Just as patients have a perspective on and can provide valuable information to the QA professional and board about quality, so can the staff of the hospital. Furthermore, the board's responsibility for quality includes the assurance of effective staffing of the hospital. The human resources plans of the hospital and its systems and policies and procedures for employee recruitment, retention, evaluation and compensation can be useful sources of quality-related information.

Staffing information such as staff turnover rates by department or area, exit interview summaries, and the effectiveness of employee recognition and reward programs are valuable for the board to occasionally receive. The staff of the hospital are most likely to know where problems in patient care exist and how they and the quality of care can be improved. The board should make certain that such a valuable information source is tapped and should have access to it.

EXTERNAL QUALITY-RELATED INFORMATION

Only after a board has a clear picture of the hospital's quality of care from internal sources, such as those reviewed throughout this chapter, should it then reference relevant external information about the hospital's quality of care. A common mistake made by many boards is to focus on external measures of the hospital's quality at the expense of more valuable internal ones. QA professionals must work with their boards to help them avoid this mistake which diverts board attention and creates false impressions of what quality is and why it is measured. This issue will be discussed in greater detail in the following chapter.

External information, in the proper perspective, can enhance the board's oversight of the hospital's quality and QA program. This type of information includes: JCAHO survey reports; professional review organization and Health Care Financing Administration data; information from private payers of health care; state regulatory review reports; national, state, and local malpractice data; and, in the near future, information from the National Practitioner Data Bank established by the Health Care Quality Improvement Act of 1985 (HCQIA, P.L. 99-660), (King, 1988).

These sources of information can help the board determine where the hospital stands relative to similar hospitals in the nation or other local hospitals. This is the most tempting use of external data sources, but not the most productive in terms of improving the quality of care.

External data sources are best used by being integrated with internal data sources to form a more complete picture of the hospital's quality. This is opposed to the common practice of using external data sources to primarily define a hospital's quality in comparison to other hospitals.

External data sources can, however, provide the board with a broader perspective on national perceptions and concerns of hospital quality of care. Further, external resources of information can also be useful to boards in the establishment goals and benchmarks for attaining higher, more desirable levels of quality.

Conclusion

Only those boards that are composed of trustees who are comfortable with the concept of quality and the purposes of QA, who are familiar with sources of information about quality of care, and who know how to and actually do analyze and act on them will be able to oversee effectively the QA program and medical staff and help assure and improve the quality of care.

This chapter has presented an overview of the information necessary to provide boards background and prepare them for their crucial monitoring and oversight of quality, QA, and medical staff credentialing. It contains a considerable amount of information which should be regarded by QA professionals as material to be used for board orientations to quality and QA, for continuing education activities, and as frequent reminders that can be included as brief introductions to the various board reports on quality.

Although this information may seem basic and too broad to be meaningful to the QA professional, it is important to stress that this represents the type of background *governance* information that boards require to perform their oversight role effectively. In fact, the material in this chapter that is too basic and broad for the QA professional will likely be regarded as too detailed and voluminous by many trustees. To effectively work with their board, the challenge to the QA professional is twofold. First, the QA professional must address the issue of quality with the board from the perspective of the board. This means to avoid forcing the perspective of the QA professional on the board regarding quality. The QA professional cannot bury the board in detail or expect the board to be aware of and support their day-to-day technical and administrative activities. Rather, the QA professional must rise above their activities and translate the clinical and management oriented information they need to do their job into *governance* information that the board needs to do its job.

The second challenge is to construct and implement a plan to effectively use the information in this and following chapters to bring the board up-to-speed on quality and to facilitate their oversight of quality and their support for the QA professional and program. The first challenge is up to the QA professional to overcome, the second challenge is addressed by the remainder of this book.

References

1. Haley RW, White JW, Culver DH, Hughes JM. "The financial incentive for hospitals to prevent nosocomial infections under the prospective payment system." *JAMA.* March 1987.
2. Haley RW. *Managing Hospital Infection Control for Cost Effectiveness.* Chicago, Ill: American Hospital Publishing, Inc.; 1986.
3. Holzman D. "The sickly side of hospital stays." *Insight.* April 18, 1988.
4. Joint Commission on Accreditation of Healthcare Organizations. *The Joint Commission Guide to Quality Assurance.* JCAHO. Chicago: 1988.
5. King MM. "After Patrick, don't forget the Health Care Quality Improvement Act." *Trustee,* November 1988.
6. Longo DR, Ciccone KR, Lord JT. *Integrated Quality Assessment: A Model for Concurrent Review.* Chicago, Ill: American Hospital Publishing, Inc.; 1989.
7. Missouri Professional Liability Insurance Association. *Risk Management Letter.* Jefferson City, MO: MPLIA; Dec 1978.

8. Orlikoff JE. "Do's and dont's for quality oversight." *Trustee.* March 1989.
9. Orlikoff JE, Vanagunas AM. *Malpractice Prevention and Liability Control for Hospitals,* 2nd ed. Chicago, Ill: American Hospital Publishing, Inc.; 1988.
10. Totten MK, Orlikoff JE. "Granting medical staff appointments and privileges: the board's key role in assuring quality care." An American Hospital Association Briefing Paper for Hospital Governing Boards. AHA, Chicago, November 1987.

CHAPTER FIVE

Developing Meaningful Board Involvement in Quality

INTRODUCTION

The first four chapters of this book have provided progressively focused information on governance and board involvement in quality as background and reference material for the QA professional. This and the remaining chapters will narrow the focus of the book further. They will concentrate on guiding the QA professional through a process of developing effective and appropriate board involvement in quality and QA, and of increasing the involvement of the QA professional with the board.

This is a process that will vary substantially from one hospital to the next, being influenced by many factors including: levels of current board involvement in quality; current general mode of board function; current board structure; size of the hospital; current relationship (if any) between the QA professional and the board; relationship between board and the medical staff; relationship between board and CEO; relationship between QA professional and CEO; level of and motivation for organizational commitment to quality; and many others.

Some QA professionals will find a receptive CEO and board while others will not. Some will have no difficulty in designing and implementing a strategy for improving board involvement in quality and QA professional involvement with the board whereas others will. However, the measurement, assurance, and demonstrated improvement of quality is an issue growing in importance in the United States and throughout the world. It is an issue that will become central to the survival and viability of hospitals and remain so far into the next decade.

This chapter lays the initial groundwork to help the QA professional develop both meaningful board involvement in quality, and meaningful QA professional involvement with the board. The first step in this process is to assess the current level of board involvement in quality as well as the level of QA professional involvement with the board.

Assessing the Board's Current Role in Quality and QA

In chapter 2, the characteristics of boards by general functional mode were presented to give the QA professional an idea of whether their board functions in the overall honorific/philanthropic, transition, or active/effective mode. It was pointed out, however, that it is possible for a board to function in one *overall* mode, but perform one of its specific roles in another mode.

For example, a board could function in the overall active/effective mode, but perform its role in quality and QA in the honorific/philanthropic mode. Much less likely, but still possible, is a board that functions in the overall honorific/philanthropic mode but performs one of its roles in the active/effective mode. If this were true for a board, the probability is great that the one role it would be actively/ effectively performing would be the financial role. It is highly doubtful that one could find a hospital board discharging its role in quality actively/effectively while functioning in the overall honorific/philanthropic mode.

So, now that a QA professional may have a general idea of what *overall* mode their board functions in, it is time to assess how their board specifically functions regarding quality and QA.

A detailed assessment checklist is presented that will allow QA professionals to assess their board's current level of functioning in quality and QA. In addition, the assessment checklist is divided into six categories to allow very specific assessments of board strengths and weaknesses in quality. These categories are: board orientation and education in quality and QA; board understanding of its role and responsibility for quality and QA; board oversight and direction of quality and QA; information flow to the board regarding quality and QA; board oversight of medical staff credentialing; and, quality and board structure.

By using the assessment checklist, a QA professional will get both an overall assessment of the board's level of functioning in quality, as well as detailed assessments of strengths and weaknesses in specific areas. This will facilitate design of a specific, tailor-made plan to improve the oversight of and involvement in quality for each individual hospital board.

There are two scoring sections for the assessment instrument. The first is the total score, which provides a score for the *overall* level of current board involvement in quality and QA. That score will place the board into one of four categories of overall involvement: Very Low (Honorific/Philanthrophic); Low (Early Transition); Medium (Advanced Transition); and, High (Active/Effective).

The second scoring section includes the six specific categories of board involvement in quality. The sum of these scores forms the total score. The score for each specific category will place the board into one of three designations relative to that category: Low, Medium, and High. The QA professional will then have a clear picture of the board's overall level of function regarding quality as well as its relative strengths and weaknesses. With this information, the

QA professional can construct and be instrumental in implementing a systematic, tailored program to improve the board's oversight of quality and consequently improve the overall quality of patient care.

Checklist for Assessing the Board's Current Role in Quality and QA

Each question in the assessment checklist is followed by a point value for a yes answer. All no answers have a point value of zero. Scoring instructions and interpretation guidelines are included at the end of the checklist.

I. Orientation and Education

1. Does each new hospital trustee receive at least one hour of orientation to the board's responsibility for quality and the hospital's QA program as part of the new trustee orientation program? *2 points*

2. Has the whole board had a continuing education session or board retreat where at least three hours were devoted to quality, QA, and/or medical staff credentialing? *2 points*

3. Have a minimum of three trustees attended the same external education program on quality and/or medical staff credentialing? *1 point*

4. If the answer to number 3 is yes, did the CEO or other senior management staff also attend? *1 point*

5. If the answer to number 3 is yes, did a medical staff leader also attend? *1 point*

6. Does the board receive periodic, brief presentations (live or video or audio tape) related to quality or medical staff credentialing? *2 points*

7. Has the board in the past two years participated in a retreat with the medical staff leadership where quality or credentialing was an agenda item? *2 points*

8. Is the board regularly given magazines, newsletters, or articles relating to governance and quality? *1 point*

CATEGORY I SCORE _____

II. Board Role and Responsibility for Quality

1. Does a written statement exist that defines the board's roles and responsibilities for quality, QA, and medical staff credentialing? *2 points*

2. If the answer to number 1 is yes, does the board reexamine and, if necessary, revise this document annually? *2 points*

3. Is there a commitment to quality in the hospital's mission statement? *1 point*

4. If the answer to number 3 is yes, is that commitment reflected or referred to in any of the following documents: board statement on role and responsibility for quality; board annual workplan; board committee workplans or charters; hospital QA plan? *1 point*

5. Does the board approve an annual workplan or calendar for itself that specifies the routine functions the board will perform and information it will receive regarding quality, QA, and medical staff credentialing? *2 points*

6. Does the board formally adopt goals for *itself* relating to discharging its role in quality, QA, and credentialing? *2 points*

7. If the answer to number 6 is yes, does the board monitor annually to determine the degree to which it has achieved those goals? *2 points*

8. If the board has conducted more than one annual self-evaluation in the past, was one of the areas identified as needing improvement board oversight or involvement in quality or credentialing? *1 point*

9. If the answer to number 8 is yes, was action taken or new activities implemented relating to board oversight of quality as a result of the self-evaluation? *2 points*

CATEGORY II SCORE _____

III. **Board Oversight and Direction of Quality and QA**

1. Does the board know what percentage of the hospital's annual budget is devoted to QA and quality-related activities? *2 points*

2. If the answer to question number 1 is yes, does the board an-

nually examine and discuss the percentage of the hospital budget devoted to quality and QA to determine if it is appropriate? *2 points*

3. If the answers to numbers 1 and 2 are yes, is the amount of the hospital's budget devoted to QA and quality equal to or greater than 2 percent? *2 points*

4. Is there a board approved definition of "quality" for the hospital? *1 point*

5. Does the board vote to approve the hospital QA plan annually? *1 point*

6. Does the board participate in or oversee (*not* simply receive reports on) the annual reappraisal of the QA program? *2 points*

7. Does the board routinely require a prospective quality plan before it makes a major decision involving new services, equipment, or methods of delivering care (examples: establishing hospital-sponsored home care program; affiliating with HMO; purchasing laser, MRI, lithotripter; establishing ambulatory surgery program; etc.)? *3 points*

8. Has the board ever taken action in response to a quality report *other* than voting to approve or accept it (for example: asking for more information; directing action; directing resources; etc.)? *1 point*

9. Does the board formally adopt and communicate specific quality-related goals or targets for the hospital (example: overall nosocomial infection rate of 4 percent; overall patient complaint rate of 2 percent; etc.)? *2 points*

CATEGORY III SCORE _____

IV. **Information Flow**

1. Does the board consider quality as an agenda item at every board meeting? *2 points*

2. Does the board receive regular quality-related reports *other* than meeting minutes of the board or hospital-wide QA committee or other quality-related committees? *2 points*

3. Do the quality-related reports to the board routinely reflect or alternate between several different perspectives of quality (such as: medical, patients' perspective, risk management, staff perspective, external perspectives)? *1 point*

4. Do the quality-related reports to the board provide trends of summarized data that are compared within the report to pre-established targets or goals (example: first quarter nosocomial infection rate = 5 percent, second quarter = 5.3 percent, third quarter = 5.5 percent, target = 4 percent)? *2 points*

5. Are quality-related reports provided to the board in graphic format with *less* than one page of explanatory narrative?   *1 point*

6. Does the board receive at least one report every six months that summarizes the frequency, severity, and causes of patient care problems, injuries, or adverse occurrences?   *2 points*

7. Does the board receive at least one report every six months that summarizes actions taken to reduce the frequency and severity of identified problems, injuries, and adverse patient occurrences, and that outlines the results of those actions?   *2 points*

8. Has the board been given information that identifies the areas in the hospital that represent the greatest risk of iatrogenic injury to patients?   *2 points*

9. Has the board received information that clearly demonstrated that the QA program had improved the quality of care?   *2 points*

CATEGORY IV SCORE _____

V. Board Oversight of Medical Staff Credentialing

1. Does the board evaluate Medical Staff Executive Committee (MEC), or equivalent body, recommendations for medical staff

appointments/reappointments against explicit criteria contained in the medical staff bylaws? *2 points*

2. Does the board evaluate MEC, or equivalent, recommendations for delineation of physician privileges in comparison to explicit criteria from the medical staff bylaws, and/or criteria established by medical staff departments? *2 points*

3. Has the board in the past two years ever requested more information from the MEC, or equivalent, before acting on a recommended candidate for appointment/reappointment to the medical staff? *1 point*

4. Based on the board's comparison of MEC recommendations against explicit criteria, has the board in the last two years voted to reverse a MEC recommendation for medical staff appointment/reappointment or privilege delineation? *1 point*

5. When acting on MEC recommendations for privilege delineation for reappointed medical staff members, does the board receive QA profile and clinical performance information on the physician? *2 points*

6. Is the credentialing process staggered such that the number of MEC appointment/reappointment and privilege delineation

recommendations received is divided up across several or all board meetings? *1 point*

CATEGORY V SCORE _____

VI. Quality and Board Structure

1. Is there a QA committee or equivalent *of* the board? *2 points*

2. Do two or more trustees regularly sit as members on the hospital-wide QA committee? *2 points*

3. Do two or more trustees regularly sit as members or invited guests on the medical staff credentials committee or the MEC? *2 points*

4. Does the hospital CEO regularly attend meetings of the board either as a member or as guest or staff? *1 point*

5. Does the senior QA professional attend those portions of the board meeting devoted to quality to present quality reports or to support written reports? *2 points*

6. Is there more than one physician member of the board? *1 point*

CATEGORY VI SCORE _____

TOTAL SCORE (Sum of all category scores) _____

SCORING SCALE

If *Total Score* is between:

A. 0–15
B. 16–38
C. 39–64
D. 65–78

Then *over-all* level of current board involvement in quality and QA is probably:

A. Very Low (Honorific/Philanthropic)
B. Low (Early Transition)
C. Medium (Late Transition)
D. High (Active/Effective)

If score for Category I—Orientation and Education—is between:

A. 0–3
B. 4–7
C. 8-12

Then rating of board orientation and education in quality and QA is:

A. Low
B. Medium
C. High

If score for Category II—Board Role and Responsibility for Quality—is between:

A. 0–4
B. 5–10
C. 11–15

Then rating of current board role and responsibility for quality and QA is:

A. Low
B. Medium
C. High

If score for Category III—Board Oversight of Quality and QA—is between:

A. 0–3
B. 4-7
C. 8-16

Then rating of current board oversight of quality and QA is:

A. Low
B. Medium
C. High

If score for Category IV—Information Flow—is between:	Then rating of current QA information flow to the board is:
A. 0–4	A. Low
B. 5-10	B. Medium
C. 11-16	C. High

If score for Category V—Board Oversight of Medical Staff Credentialing—is between:	Then rating of current board oversight of credentialing is:
A. 0–2	A. Low
B. 3-5	B. Medium
C. 6-9	C. High

If score for Category VI—Quality and Board Structure—is between:	Then rating of board structure for quality is:
A. 0–3	A. Low
B. 4-7	B. Medium
C. 8-10	C. High

After reviewing the assessment instrument it will become obvious to the QA professional that in most situations the CEO or other senior executive will need to be consulted and enlisted in order to gather the answers to many of the assessment questions. There are several ways the QA professional can approach this.

The QA professional can approach the CEO or other senior executive and explain the purpose of the assessment instrument and request permission and assistance in completing it. If the request is met by skepticism, two arguments can be used to counter the skepticism.

The first is the direct and obvious one, that it is necessary to accurately and honestly assess the board's current role and involvement in QA in order to strengthen and improve it. Further, by strengthening and improving the board's role in QA, the hospital will benefit by improving the QA functions and the quality of care, and reducing the chances of being cited for failing to comply with JCAHO standards and other regulatory requirements.

The second argument is that the assessment instrument can be used as the basis for conducting a board self-evaluation, which is required by the JCAHO. The assessment instrument can be used as is, or be modified by the board to conduct a self-evaluation that focuses on the board's role in quality. (This is also a good way of introducing and involving the QA professional with the board and beginning the important partnership for quality between them.)

In the event that the CEO or senior executive is not receptive to the idea and is unconvinced by these arguments, the QA professional must confront the fact that access to and involvement with the board regarding quality is currently limited. This does not mean that it will always be so, and the QA professional can take several actions to change the situation and assess and improve the board's involvement in QA and quality.

In this situation, the first step is for the QA professional to complete as much of the assessment instrument as possible alone. Then, acting on the results of the assessment, the QA professional can begin to effect change in the board's involvement in quality by changing the quality-related reports to the board. By making the reports *educational* as well as informational, the QA professional can begin to change the board's perspective on and interest in quality and QA. The reporting of quality-related information to the board is discussed in detail in the next chapter.

Should the QA professional be in an environment where changing the formats of the QA reports to the board or board QA committee is not possible or permitted, then the options are severely limited. One option of questionable value is for the QA professional to consult directly with a member of the board regarding the matter. This type of end-run approach, however, rarely results in positive outcomes and is generally ill-advised.

A QA professional unlucky enough to be in an environment where access and reporting to the board are prohibited or severely controlled, and where the CEO does not recognize the critical importance of meaningful board involvement in quality is in an unfortunate situation indeed. In this type of extreme and, thankfully, rare situation there is really very little that the QA professional can do to change the circumstances that limit access and involvement with the board. The only options are seeking new employment or waiting for a significant change in the organization's structure and culture. For those QA professionals in such a situation there is little solace in recognizing that the situation does not bode well for the quality of care provided, the QA program, or for the very future of the hospital itself.

Assessing the QA Professional's Involvement with the Board

Once the current level of board involvement in QA and quality has been assessed, one other step remains before an action plan for improving board involvement can be developed and implemented. This step involves the QA professional assessing the current involvement and relationship with the board. This is advisable because as the board's level of interest and involvement in quality and QA increases, it is an opportune time to improve the board's perspective of and relationship with the QA professional. It is possible, but certainly not desirable for the QA professional, for a board to become more interested and involved in quality without becoming more interested and involved with the QA professional. It is incumbent upon the QA professional to positively manage this change.

In order to do this effectively, the QA professional must first assess the relationship with the board generally, and attempt to specifically determine how the board perceives the QA professional.

The current general relationship between the board and the QA professional has been framed by such variables as: QA reports to the board; past QA professional presentations to the board; staffing of any board QA-related committees; and, the circumstances and situations where the QA professional has been involved with or brought to the attention of the board. This last point is worth stressing.

If the only time the board is aware of the QA professional is when the hospital is preparing for a JCAHO survey, responding to the contingencies of a recent survey, or preparing for a focused review, then the board is likely to associate the QA professional, and indeed QA itself, with JCAHO compliance and accreditation. This is clearly a limited and therefore limiting view of the role and capabilities of the QA professional. A board that holds this view is unlikely to regard the QA professional as a key resource in the new push for quality. Unlikely, that is, unless the QA professional is able to manage both the change in the board's involvement in quality *and* a change in the board's perception of the QA professional.

To do that effectively, the QA professional must give some thought to current perceptions by the board, as well as the CEO and senior management, and how those perceptions can be changed and enhanced as the board becomes more focused on quality. The QA professional must determine how they wish to be perceived by the board and then structure their activities with the board to create that perception. This individual process will vary from hospital to hospital depending upon such variables as current levels of board involvement in quality; the current relationship between the board and QA professional; the ability of the QA professional to change the parameters of the job; and many others.

Regardless of the individuality of the process, the key to success begins with the establishment of a goal of how the QA professional wants to be perceived and what role the QA professional would like to have in the hospital's quest for quality. Then, as the QA professional begins to increase the board's involvement in and understanding of quality, they can also structure their involvement with the board so as to change the way they are perceived, and to change and elevate their role within the hospital.

The essence of the transformation of the perception and status of the QA professional within the hospital rests upon a simple premise: the QA professional is the hospital's resident expert in quality and its many and varied facets. This means that the QA professional must structure reports to and interactions with the board in such a way as to be perceived as an *expert* in quality who *directs* the hospital's QA program and quest for quality. This is very different from the QA professional who is, or is perceived as being, the person who *staffs* the QA functions and committees and who primarily responds to external requirements.

Clearly, this is a challenging transformation to effect. It is one that will not be easy or quick to bring about. It is, however, made much easier and indeed possible by the very fact that the board will be undergoing a transformation of its own as it increases and becomes comfortable with its involvement and oversight of quality. The key for the QA professional is to seize the opportunity and to directly link *their* transformation to that of the board's. Thus, the QA professional should closely link the process of enhancing the board's oversight of quality with the process of enhancing their own role and status.

DEVELOPING AN ACTION PLAN

At this point the assessment of the board's current role in quality and QA has been completed, and the QA professional has given serious thought to the current and desired role with the board and within the hospital. It is now time to use the results of these two assessments to develop and implement an action plan. The purpose of the action plan is to guide the systematic improvement of the board's involvement in and oversight of quality and QA, and to enhance the role of the QA professional with the board and within the hospital.

The results of the assessment of the board's current involvement in quality and QA (hereafter referred to as the board assessment) will provide the basic framework for the action plan. Thus, it is necessary to examine the results of the board assessment in detail.

First, the *Total Score* of the board assessment should be considered. The lower the total score of the board, the more work that will be necessary to develop and implement an action plan to improve the board's role in quality and QA. A board that has a total score of between 0–15 is in the honorific/philanthropic stage of quality oversight, and is also likely to be performing several of its other functions in the honorific/philanthropic mode. This type of board poses both significant challenges as well as opportunities for the QA professional.

The challenges posed by this type of board relate to the fact that it faces a long journey to attain the active/effective mode of quality oversight. This also presents the opportunity for the QA professional to guide the board during their evolution, and so to be viewed as an educator and consultant to the board, and as the resident expert in quality.

After the total score of the board assessment is determined and the general picture of the board's involvement in quality and QA is known, the next step is to examine the scores for each *category* of the assessment instrument. This will identify for the QA professional the relative strengths and weaknesses of the board in quality and QA, and will provide the framework for the action plan.

In the case of a board that scores in either the Very Low (honorific/philanthropic, score 0–15) or Low (early transition, score 16–38) level of overall current involvement in quality and QA, the scores of the individual categories of the board assessment will be less meaningful in the development of the action plan than will those of boards that have higher total assessment scores. In this situation, the action plan will need to be constructed to cover the development and improvement of all of the aspects of effective board involvement and oversight in quality and QA. Here, the results of the board assessment categories will be most useful in identifying current strengths to build an action plan around.

In the case of the board that generates a total score of Medium (late transition, score 39–64), or High (active/effective, score 65–78), the action plan will focus more on fine tuning and maintaining the board's involvement and oversight of quality and QA than on striving to achieve it. Here, the foundation for effective board oversight of quality and QA should be in place, and the action plan should relate to maintaining the integrity of that foundation and to strengthening the board's understanding and effective discharge of its responsibilities. In this situation, the categorical results of the board assessment will provide the direct framework for the development and implementation of the action plan by identifying the current strengths and weaknesses of the board's involvement in and oversight of quality and QA.

THE BASICS OF THE ACTION PLAN

Most action plans for developing or enhancing board oversight of quality and QA will have several basic themes in common. These basic themes include: orientation and education; the determination and affirmation of specific and relative roles and responsibilities between the board, medical staff, management, and QA professional regarding quality and QA; the establishment of quality indicators for the board along with thresholds for action; the development and refinement of meaningful quality reports to the board;

the examination and refinement of quality-related information flow to the board; the design and implementation of effective organizationl structures for the board's involvement in and oversight of quality and QA; and, the integration of quality indicators and QA into an improved process for board consideration and action on medical staff credentialing.

For boards that generate a high total score and uniformly high categorical scores in the board assessment the basic themes mentioned above may have already been addressed and the board may be functioning well regarding quality and QA oversight. In these situations, an action plan is still useful for maximizing effective and efficient board oversight of and involvement in quality and QA, and for enhancing the role of the QA professional.

For these relatively advanced boards, the action plan will focus on more advanced themes such as: enhancing and refining the basic themes; integration of quality assurance, risk management, and utilization management information and oversight at the board level; integration of *prospective* quality assurance and risk management into all relevant board decisions and actions (pages 48 and 49); commitment to the development of an organizational culture for quality and the implementation of total quality improvement programs; development and refinement of performance-based medical staff credentialing systems; integration of quality into the board's and hospital's mission evaluation and strategic planning activities; enhancing the strategic partnership for quality between board, medical staff, management, and QA professional; and ongoing and effective orientation and education.

One of the common *and* advanced themes in any action plan for improved board involvement in and oversight of quality is orientation and education. It is important to stress that the education of the board and its members will be an ongoing process which does not end with an orientation program for new trustees or a few members of the board attending an external seminar related to quality. Further, education regarding the board's role in quality and QA should not be limited to only the board. Rather, it should include the medical staff and focus on such issues as the board's ultimate responsibility for quality and how this operationally translates into a practical distinction of relative roles and responsibilities between board and medical staff. It should also include hospital staff and employees to clearly demonstrate the top-level commitment to quality in the organization.

Orientation and education will be the cornerstone of the action plan for boards that scored medium to low on the board assessment, and will also be a key element of the action plan for those boards that are relatively advanced regarding quality and QA.

Education is the key to board members understanding *why* the board is ultimately responsible for quality *and* the various ways of effectively discharging this key responsibility. An action plan that ignores or places minimal emphasis upon orientation and continuing education is likely to fail. Boards will be unable to discharge their responsibilities for quality unless they clearly know what those responsibilities are, and why the board is charged with those responsibilities. Much of the content material for education on this subject was covered in chapters 3 and 4.

Education is also a key technique for the QA professional to enhance their role with the board and within the hospital. By designing and participating in the quality-related orientation of new trustees, and in the continuing education of the board regarding quality, the QA professional will position themself to be perceived as an educator and consultant to the board, as well as the resident expert in quality. From these contacts will flow other levels of involvement with the board. Greater board emphasis upon quality and QA, increased reliance upon the QA professional, and an enhanced role for the QA professional are a likely result.

All effective action plans will have several common characteristics necessary for success. These include:

- A goal or set of goals for the action plan is established, and specific goals and objectives for each of the different components of the action plan are established.

- The action plan is tailored to address the specific characteristics of each individual board (hence the need for some type of board assessment prior to the development and implementation of the action plan).

- The action plan addresses both potential changes in board function and structure regarding quality and QA, but considers function *before* structure and examines structural changes (i.e., the establishment of a board QA committee, or the modification of existing committee structures or composition) for the purpose of *supporting or facilitating* effective functional changes.

- The action plan addresses both board monitoring and oversight of *quality* as well as monitoring and oversight for the effectiveness and integrity of the *QA program* and various QA functions, especially those of the medical staff.

- The action plan specifically addresses the establishment of quality indicators and appropriate and effective reporting formats for the board.

- Time frames for the implementation of the action plan and its components are established.

- An evaluation of the action plan is built into the action plan. The evaluation should assess the outcomes, strengths and weaknesses of the design and implementation of the action plan as compared to the goals and objectives of the action plan and should result in modifications in function and structure as appropriate.

- The board approves the action plan and supports its implementation.

The absence of the above characteristics from an action plan lessen the chances that the plan can be implemented successfully. The fewer of the above characteristics that are present, the greater the risk of a failed effort.

The last characteristic of an action plan, that the board approves and supports the plan and its implementation, bears further comment. This characteristic is key to assuring board buy-in to the plan, and hence to increasing the probability of the plans success. It is not always possible to obtain the board's approval, however.

In cases where the board is in the honorific/ philanthropic mode, it may not see a need to improve its oversight of quality and QA. Conversely, boards that are higher up on the scale and are currently addressing quality, albeit not in the most appropriate or effective manner, may also not see a need to focus on improving quality and QA oversight. Other factors may preclude the board's approval of the plan, such as a CEO who is not committed to the process and refuses to support the development of the plan or allow it to come before the board.

In situations where the board does not see the need for such a plan but the CEO does, the inability to get board approval of the plan should not prevent its development and implementation. It will obviously require, however, that the plan be modified to focus on such initial goals as educating the board about *why* it needs to be concerned with improving its oversight of quality and QA, as well as how to go about subtly doing that without explicit board involvement in the process.

The lack of explicit board involvement in or of tacit board approval of the process should therefore not preclude the development or implementation of an action plan for improving the board's involvement in and oversight of quality and QA. In fact, it speaks to the critical need for such a plan.

In the unfortunate situation where the CEO does not support the development of the action plan and refuses to allow it before the board, the QA professional can still develop and implement an action plan of their own. This plan should include as many of the key characteristics covered previously as possible.

In this situation, the plan developed and implemented primarily by the QA professional will focus on such goals as changing the attitudes of the board and CEO, and on facilitating improved board involvement in quality and QA through implicit means. Usually, the main vehicles of achieving this are the QA reports and presentations to the board, reporting to and staffing of any board QA committee, and providing education or educational opportunities and background literature that is *appropriate for trustees* to the board. This approach is clearly a more difficult one for the QA professional than when the board and CEO are supportive of the plan, but it is nonetheless worth pursuing. Indeed, one of the primary goals of this approach is generating support from the board and CEO. Consequently, this strategy may take a fair amount of time to come to fruition, so a QA professional's action plan in this situation should be built upon realistic timeframes and implemented in a politically delicate manner.

A sample action plan is presented here. Although the sample action plan is geared toward a board that scored Low (Early Transition, score 16–38) on the assessment instrument, it is important to stress that other boards with similar *overall* scores may require very different action plans. This is due to the need to tailor the action plan to the specific characteristics of each individual board. Different boards with the same overall scores may have very different strengths and weaknesses as reflected in very different categorical scores on the assessment instrument.

Another point worth stressing is that the timeframes for implementing the action plan will vary significantly as a function of individual board and hospital organizational structure and political climate. To give one simple example of this, some boards meet monthly while others meet quarterly. The timetables for implementation of the action plans for two boards with different meeting schedules will clearly be different, even if the scores and other circumstances of the boards are similar.

Regardless of the board or the intricacies of the action plan, the QA professional must take care to make the timetables for the implementation of the action plan as realistic as possible. The transformation of a board from minimal involvement, oversight, and understanding of quality and QA to a board that is comfortable with, appropriately involved, and effectively overseeing quality and QA is not a quick journey or a simple task. The timetables of the action plan should reflect this by being ample and flexible.

SAMPLE ACTION PLAN FOR A BOARD SCORING LOW (Early Transition, 16–38) ON ASSESSMENT

Overall Action Plan Goals
To increase the board's understanding of its responsibilities for quality and quality assurance; to improve the board's understanding and oversight of the hospital's quality assurance program, the hospital's quality of care, and the medical staff credentialing process; to develop meaningful and understandable quality indicators and reporting formats for the board; to make quality a priority for the board and the hospital; to develop strategic approaches to expressing and achieving the board's quality priority.
Time Frame: 18 Months.

I. *Board Orientation and Education*
   A. Devote all or majority of board retreat session to review of board's responsibility for quality, review of quality and QA, and the board's role in overseeing quality, QA, and medical staff credentialing. Also review the hospital's QA program and current indicators of quality. Use external consultant or faculty with input and involvement from QA professional.

Time Frame: Develop, schedule, and conduct retreat—Months 1–3.

A (Alternate). If board retreat dedicated to QA not possible, design and conduct series of half-hour quality and QA orientations to be conducted at the beginning of the next six board meetings. Content the same as under A. above. Use publications and videotapes designed for trustee audience on quality and related topics. Faculty: QA professional, CEO, VPs for nursing and medical affairs, chief of staff.
Time Frame: Develop and schedule sessions—Months 1–2; Conduct sessions—Months 2–8.

B. Develop a minimum of three one-hour QA education sessions as QA module of hospital's new trustee orientation program.
Time Frame: Develop sessions—Months 8–10; Conduct sessions—at least once annually thereafter during orientation of new board members.

C. Develop ongoing continuing education strategy for board: develop briefing sessions; screen and select reference materials; select video tapes and articles; review and recommend external education programs; develop format for summarizing legal and regulatory developments as part of regular quality-related reports to the board.
Time Frame: Develop strategy—Months 9–11; Begin ongoing implementation—Month 11 and thereafter.

D. Develop and conduct educational session for medical staff on the board's role and responsibilities for quality, QA, and medical staff credentialing.
Time Frame: Months 3–5.

II. *Developing and Enhancing the Board's Role in Quality*
   A. Securing Board Commitment to Quality
      1. Following its presentation, board approves action plan.
         Time Frame: Month 1.
      2. Board approves policy statement on quality, and its

role in it. Statement defines quality for the hospital, or the various perspectives that are related to quality.
Time Frame: Months 10–12.
3. Relative roles of the medical staff, management, QA professional, and board regarding quality, QA, and medical staff credentialing are defined and incorporated into hospital QA plan and board policies.
Time Frame: Months 10–12.

B. Board Oversight of Quality
  1. Develop and implement quality indicators for the board.
  Time Frame: Months 9–11.
  2. Develop and implement indicators of the function and effectiveness of the QA program and medical staff QA monitoring activities.
  Time Frame: Months 12–14.
  3. Develop and implement quality indicator targets or thresholds for board action.
  Time Frame: Months 14–16.
  4. Determine amount of annual hospital budget devoted to quality and QA, integrate into board's annual budget review and approval process.
  Time Frame: Months 16–18, and annually thereafter.
  5. Develop annual board workplan for quality, outlining the board's quality indicators, focus, and goals for quality, QA, and credentialing for the next year.
  Time Frame: Develop workplan, board approves workplan—Months 16–18.

C. Board Oversight of Medical Staff Credentialing
  1. Research and list all current decision criteria that are, or are supposed to be, used by the medical staff to make medical staff appointment, reappointment, privilege delineation, and privilege renewal and restriction recommendations. Review these criteria with the board.
  Time Frame: Months 15–17.
  2. Review relative roles and responsibilities for medical staff credentialing, with emphasis on the board's role, with the board.
  Time Frame: Months 15–17.

Note: Other action items for board oversight of medical staff credentialing are covered under following two sections.

III. *Quality Information Flow and Reporting*
   A. Develop and implement calendar for the rotational reporting of board quality indicators, quality-related and QA program information to the board; specify areas, departments, committees, etc., that will provide information to board.
   Time Frame: Months 16–18.

   B. Develop standardized format for reporting quality indicators and targets or action thresholds to board. Board reports should demonstrate *trends,* and be in graphic format whenever possible.
   Time Frame: Months 16–18.

   C. Develop format for reporting medical staff recommendations for credentialing decisions to board that allows the board to compare the medical staff recommendations against the applicant's compliance to existing explicit criteria.
   Time Frame: Months 16–18.

IV. *Quality and Board Structure*
   A. Develop pros and cons of establishing board QA committee, establish potential charge/purpose of committee, and outline reporting relationships. Present to board for consideration and possible approval.
   Time Frame: Months 17–18.

   A (Alternate). If board QA committee or equivalent exists, develop refined statement of purpose and outline of reporting relationships.
   Time Frame: Months 15–17.

   B. Board approves recommendation that quality be an agenda item at every board meeting.
   Time Frame: Months 10–12 (with approval of policy

statement in II.A.2), or Months 14–16 if policy statement not approved, or approved without agenda commitment.

C. Develop recommendations for board consideration and implementation relating to trustees as members of the hospital-wide QA committee, Medical Executive Committee, and other quality and credentialing related committees.
Time Frame: Months 17–18.

D. Review and refine reporting relationships to board regarding quality and QA.
Time Frame: Months 15–17.

V. *Action Plan Evaluation*

A. Conduct reassessment of board involvement in quality using same instrument as was used for the initial assessment as the foundation for the reassessment.
Time Frame: Month 18 and thereafter.

B. Conduct evaluation of the strengths and weaknesses of the results of the implementation of all components of the action plan.
Time Frame: Month 18 and thereafter.

C. Based upon the results of A. and B. above, design and implement modifications and refinements to action plan and to all related aspects of board involvement in and oversight of quality.

A board that scored higher on the assessment instrument than 16–38 would require an action plan that reflected more sophisticated issues and strategies. Such issues and strategies would also be considered and implemented as an initially low scoring board evolved in its understanding, involvement, and oversight of quality and QA as a result of the successful implementation of the action plan.

Such an advanced action plan might include the following issues and strategies:

- education and orientation for members of the board QA committee, or its equivalent;
- the integration of severity-adjusted measures of quality into the board quality reports (education for the board on this topic would precede implementation);
- the development of a performance-based, Level 1 and Level 2 criteria system for medical staff credentialing (see chapter 4);
- education and orientation of medical department heads regarding the establishment and use of credentialing criteria;
- designing and conducting board, medical staff leadership, and management retreats on strengthening the partnership for quality;
- the development and implementation of continuous quality improvement programs (such as those based on the methods of Demming, Juran, and other models);
- the integration of the hospital's culture for quality with external requirements for QA and monitoring activities;
- the integration of *prospective* QA and risk management into all relevant board decisions;
- building CEO commitment to quality and QA through such methods as developing CEO evaluation criteria that relate to quality;
- the establishment of a position of "Chief Quality Officer" that, as does the Chief Financial Officer, reports directly to the CEO;
- the establishment of annual "quality goals" for the hospital and the aggressive pursuit of achieving those goals.

As boards become more involved in quality, the action plans to guide each board through further development will become more focused and remarkably board-specific. This will result from the different perspectives, emphasis, and approaches to quality that different boards will develop as they integrate quality and QA into their accepted and ongoing governance functions.

CONCLUSION

The development of meaningful board involvement in and oversight of quality is not a task that is best left to circumstance or to halfhearted planning or execution. It is a process that requires careful planning, delicate implementation, and continual refinement.

The process consists of assessing current board understanding of and involvement in quality, developing an action plan based on that assessment, implementing the action plan, evaluating the effectiveness of the action plan, and beginning the process over again. Each time one cycle is completed the board will have advanced in its understanding of and commitment to quality and QA. The next process cycle will be more focused, more advanced, and will continue the journey of the board on the road to quality.

CHAPTER SIX

Providing the Board with Meaningful Quality-Related Information

INTRODUCTION

The next step in the process of developing meaningful board involvement in quality and QA is extremely important. It concerns the issue of defining the types of QA and quality-related information that will go to the board, and the methods and formats in which that information will be presented to the board.

Both the information itself and the way it is communicated will have a significant impact on the board. It will influence the appropriateness of board involvement in quality and frame the effectiveness of board oversight of quality and QA. Because of this, careful consideration by both the QA professional and the board should be given to the substance and format of quality reports to the board.

GOVERNANCE INFORMATION VERSUS MANAGEMENT AND CLINICAL INFORMATION

It has hopefully been made clear in earlier chapters that the board has the ultimate responsibility for the quality of patient care and the QA program of the hospital. This does not mean that the board *manages* the hospital or *actually pro-*

vides medical or clinical care. The vast majority of hospital CEOs, physicians, trustees, and QA professionals clearly agree with this. Yet, the information that flows to most boards regarding quality and QA is often indistinguishable from the information that is seen by management, QA professionals, or the medical staff.

This is a critically important point. The information reviewed and acted upon by any group within the hospital *must be appropriate to the role and function of that group.* This is because the level, detail, and format of information provided to any group will actually structure and determine the way the group analyzes and acts upon the information.

Thus, if a board is constantly provided with management information, it will likely respond by performing management functions. Similarly, if a board is constantly provided with clinical QA data that is appropriate to medical, nursing, ancillary, or QA departments, the board may be induced to perform functions appropriate to clinical or QA departments.

Conversely, many boards are provided with quality and QA information that is not appropriate to management, the medical or clinical staffs, or QA. Unfortunately, this information is usually inappropriate for the board as well as it is too broad and unfocused to be meaningful as governance information. Here, too, the information will structure the board function regarding quality and QA. Broad, unfocused quality information results in unfocused ill-defined, and ineffective board involvement in quality and QA.

Sending QA information to the board without first defining the role and function of the board regarding QA is a classic example of the tail wagging the dog. In this all-too-common situation, the role of the board is defined and structured *by the information it receives.* In fact, the proper way it should be done is just the opposite. The information sent to the board should support and reinforce the board's predefined role and facilitate the effective performance of the board's recognized and accepted functions regarding quality and QA. This type of information is *governance information.*

It is therefore important to recognize the distinctions between *management or clinical information* and *governance information.* This distinction is important for all types of information that goes to the board, but it is absolutely critical for QA and quality-related information. This is much more than a theoretical concept as it will directly influence the content, format, and frequency of quality-related information that goes to the board. This in turn will directly influence, if not control, the board's involvement in and oversight of quality and QA.

The determination of what kinds of quality-related information and reporting formats are *appropriate* to the board, and the ongoing reexamination of this question and refinement of report content and formats is the essence of providing meaningful quality-related information to the board. It is also a key element in facilitating appropriate and effective board involvement in and oversight of quality and QA.

Typical Weaknesses of Quality and QA Reports to the Board

Before addressing the issue of what quality-related reports to the board *should* be, it is instructive to review what they *should not* be.

First, most current QA reports to the board do not flow from or support the explicitly defined role of the board regarding quality and QA. While this may be due to the fact that few boards have actually developed and approved a defined statement of their role or function regarding quality or QA, it nevertheless results in quality reports that are management or clinically oriented. Most current quality reports to the board do not provide governance information.

A second common problem related to the first is that there are no guidelines regarding what information should be reported to the board, or how it should be reported. In this common situation, the result is that the board is *directed by* the information it receives, and thus is directed by the QA program. The way it should be is that *the board directs or specifies* the information it wants to receive, and thus the board will be able to direct or oversee the QA program and quality-related functions of the hospital.

Another common problem with quality reports to the board is that the reports provide *data* and not *information*. In this context, data are raw numbers of simple statistics. Information, on the other hand, is analysis that clearly demonstrates what the data means. The provision of data, not information, to the board is due to several factors.

One such factor is the provision of *numerator* data to the board without *denominator* data. An example of this is the report that informs the board that the hospital nosocomial infection rate for the month of October is 6.5 percent. Is that rate high or low? Is it good or bad? What does it mean? The fact is, that piece of numerator data is meaningless unless compared to such denominator data as: past trends in infection rates, relevant variances in occupancy rates, and other pertinent information such as recent changes in the definition of nosocomial infections or in monitoring techniques.

Other common examples of this type of "numerator data only" reporting to the board are: monthly incident report numbers that are not broken down by category of patient injury or location, or are not compared to past trends in incidents; the number of medication errors last quarter compared to nothing; the number of medical records that were screened last month with no results of those screening functions provided; the number of surgical cases that were reviewed last month with no analysis of how many variations were justified by physician reviewers; and on and on.

Presenting the board with numerator data without denominator data is a waste of the board's time. This is because numerator data alone is meaningless to the board (as well as to anyone else) without something to compare it to—the denominator data. Because numerator data without denominator data precludes comparative analysis, no meaningful conclusions can be drawn from it or appropriate actions taken.

Consequently, the provision of "numerator data only" to the board results in one of two equally inappropriate board outcomes. The first is that the board accepts the data without asking important questions, or is unable to discern important problems in quality or the QA process which should be corrected. The second is that the board actually responds to the data, taking inappropriate action or directing resources toward the solution of nonexistent or inconsequential problems.

Another factor that contributes to the provision of "numerator data only" to the board is the tendency to communicate *cross-sectional* data as opposed to *longitudinal* information. Cross-sectional data are disjointed slices of data framed by finite and arbitrary periods of time.

The previous example of the 6.5 percent nosocomial rate in the month of October demonstrates this. By itself, it means nothing to the board or anyone else. When presented in relation to the 6.0 percent nosocomial rate in September, it takes on some, albeit limited, meaning. When compared to the steadily increasing nosocomial infection rate of the last year (increasing from 2.5 percent one year ago to 6.5 percent one month ago in October), however, it has a clear and disturbing meaning that the board can and should appropriately address.

Providing longitudinal information about quality and QA to boards is even more critical in small hospitals with small medical staffs or low numbers of admissions or certain procedures. In these situations, cross-sectional information is most likely meaningless, as trends of quality problems will be masked by small sample sizes. Here, information to the board *must* be longitudinally based, should identify trends, and should always include denominator data.

Clearly, boards require governance information, and not "numerator data only" or cross-sectional data. Unfortunately, the provision of "numerator data only" in cross-sectional frames of reference is a common if unfortunate method of providing boards QA and quality-related information. These common reporting methods provide the board with limited information and consequently blunt the effectiveness or appropriateness of the board's response to the reports. Thus, the content and format of the QA reports themselves actually preclude many boards from effective involvement in and oversight of QA and quality.

Another common problem with quality reporting to boards is the simple submission of sets of meeting minutes to the board as quality and QA reports. These often include the minutes of the following committees: the hospital-wide QA committee; the credentials committee; the Medical Executive Committee; the joint conference committee; or the board QA committee.

While providing the board with minutes of various committees as *background* information may be appropriate, it is inappropriate and unacceptable to provide boards with minutes of committees for the exclusive purpose of quality or QA reporting. This is true for several reasons.

First, minutes are simply the records of meetings. As such, they provide information relevant primarily to the group that was meeting, or to similar or related groups. Thus, a medical staff committee meeting produces minutes that contain information appropriate to the medical staff, not the board. In other words, most minutes contain information that is *not* governance information.

Furthermore, minutes of meetings generally present cross-sectional information and often contain only numerator data. Thus, it is extremely difficult for a board to discern trends from minutes, or to get a realistic assessment of anything other than how a specific committee functions (with some minutes, even that is not possible).

In addition, most meeting minutes are dry and make

for tiresome reading. Consequently, many trustees *do not read the minutes of meetings they did not participate in.* For all of these reasons, the minutes of meetings *should not* be used as the primary vehicle for reporting quality and QA information to the board.

Although most of the examples given of inappropriate and ineffective reporting of quality and QA information to the board have related to *not enough meaningful information,* another less common weakness of some board quality reports is *too much information.* The reporting of quality and QA information to the board is important, but it is possible to take it to the extreme and provide the board with excessive amounts of data, detail, or even governance information.

In situations where this occurs, the board often finds itself buried in data and information. Boards can respond to this in one of several ways. A board can become groggy and inured in response to the stultifying amounts and detail of the information provided and gradually "tune out" of quality and QA.

Some boards react to being buried by excessive amounts of information by taking it as incorrect evidence that the hospital's QA program is functioning well and that the quality of care is good. In this situation, the board often gets lulled into a false sense of security about quality and QA, believing that "its being taken care of—just look at all the paper and reports we get about it." The board does not respond to these information overloads, because it has no way of sifting through the piles of paper to find meaningful information about quality and QA.

Still another board response to excessive quality and QA reporting is to become *inappropriately over-involved* in quality and QA. A small number of boards become obsessive about quality and QA when provided with excessive quality information that is more management and medically oriented than it is governance oriented. In these situations, the board begins to perform quality- and QA-related functions that are more appropriate to management, the medical staff, and the QA professional. This causes many problems within the hospital including conflicts between the board and the management and medical staff, and impedes the effort to provide high quality care that is continuously examined and improved.

Another common flaw in the reporting of quality-related and QA information to boards is the exclusive reporting of QA process information. This is information that demonstrates that the QA program and its many processes are indeed functioning.

Unfortunately, when this is the *only* information about quality that is reported to the board, it usually demonstrates that certain paper-pushing QA processes are being conducted. It does not, however, indicate *how well* or *effectively* these processes are being performed.

The exclusive reporting of QA process information to the board is dangerous in that it has no meaning unless compared to the *outcomes* of the QA processes. Are all these QA processes and functions and reviews actually resulting in improved quality care? If that question cannot be easily answered by a board, then the board is being provided with useless information. Furthermore, although the board needs to know how the QA processes are functioning, the excessive reporting of QA process information tends to mask information about outcome-oriented quality of care. As with QA itself, the board should be more focused on outcome than on process.

Even when outcome information is combined with QA process information to the board, a common weakness in board reporting is that relatively unimportant issues are reported to the board at the expense of important quality issues or problems. Too often, quality reports to the board routinely convey the impression that "all is well," or "the quality of care is fine, don't worry," when in fact there *are* problems in quality or issues that require immediate attention.

Often, reports will detail the identification and resolution of a minor problem. The problem with this approach is that it frequently, perhaps intentionally, *hides significant quality problems or issues from the board.* Here, a board can receive seemingly significant quality and QA information and still not have a real sense of the hospital's quality of care or effectiveness of the QA program.

Another common problem with quality reports to the board is that the presentation formats are unclear or disorganized. Examples of this can be seen in long narrative reports when a simple graph would give the same information much more clearly and effectively; or in extensive tables of data that would take hours to analyze.

Barry A. Passet, the President of the Greater Southeast Community Hospital Foundation, sums up many of the inherent problems in reporting quality and QA information to the board:

> Many boards today are caught up in reviewing the accreditation-driven, paper *process* of quality assurance: Medical records pulls charts. Physicians review charts. They plot trends and make severity adjustments. They do reports. Medical departments and committees process the reports. The Medical Executive Committee reviews the reports. The Board's Quality Assurance Committee studies the reports. The Board reviews the reports.
>
> The bottom line: Nothing changes. That is simply wasted effort. It is costly. It turns people off and inculcates them with the certain knowledge that quality assurance is a waste (Quality Letter for Healthcare Leaders, July 1989).

If the quality reporting process to the board, both in terms of content and format, creates the impression among board members that "quality assurance is a waste," how will the board then view the QA professional?

Here is a summary of the common flaws and weaknesses found in quality reports to hospital boards. The list is fairly extensive and should be regarded by QA professionals as the way *not to do it.*

Common Flaws and Weaknesses in Quality Reports to Hospital Boards

1. Reports do not flow from or support the explicitly defined role of the board regarding quality and QA.

2. There are no guidelines regarding what information should be reported to the board or how it should be reported.

3. Reports do not contain *governance information.* Rather, they present management or clinical information.

4. Reports provide *data* but not *information.*

5. Reports provide numerator data only, without denominator data to facilitate comparative analysis and understanding by the board.

6. Reports present cross-sectional data only, not longitudinal information that clearly demonstrates trends.

7. Meeting minutes are used as the exclusive vehicle for providing quality and QA reports to the board.

8. Too much data or information is presented in the reports.

9. The reports provide information only on the QA process, with no information on outcomes or on quality of care.

10. Reports present or address only insignificant problems or issues and hide significant QA and quality issues from the board.

11. Ineffective report formats blunt the board's understanding of important information.

Clearly, the format and content of quality reports to the board are crucial. They directly, if not entirely, influence how the board regards the function of QA, and how effectively they are involved in and oversee QA and the hospital's quality of care. They form the board's assessment of the hospital's quality of care. Further, they are instrumental in forming the board's opinion of and regard for the QA professional and the work they do.

A Process for Developing QA Information for the Board

In order to avoid the common pitfalls of reporting quality information to the board, and to facilitate the board's *direction and oversight* of quality and QA as opposed to the board *being directed by* the QA and quality information it receives, an organized process is required. This organized process will not only structure and define the information the board will receive about quality and QA, but it will also formalize the board's role in overseeing the QA program and the hospital's quality of care.

Although the process may vary somewhat from hospital to hospital, to be successful all will share several common characteristics. These common characteristics will include: that the process is formalized; that it is developed with input from the board and the QA professional; and, that it is regularly evaluated and refined.

Such a formal process for developing quality and QA information to the board is presented here.

1. Develop a statement of the board's role in QA and quality.

2. Develop guidelines for the type of internal and external QA and quality-related information (both content and format) that will go to the board.

3. Develop and select a series of quality indicators for the board.

4. Determine threshold or target levels for the quality indicators.
5. Develop report formats for reporting the quality indicators and threshold levels to the board.
6. Define the other (non-quality indicator) QA and quality-related information that will go to the board.
7. Develop a calendar for quality reporting to the board.
8. Monitor and evaluate effectiveness of quality indicators, refine them.
9. Monitor and evaluate appropriateness of quality indicator threshold levels, modify them as appropriate.
10. Evaluate the entire process and refine as appropriate.

Several of the above points will have been addressed by the action plan discussed in chapter 5. The remaining points, once developed, should then be incorporated into or appended to the action plan as well.

The remainder of this chapter will address the key points of the above process in detail.

Develop a Statement of the Board's Role in Quality and QA

As discussed in earlier chapters, there are several purposes for developing a defined statement of the board's role in quality and QA. In addition to clarifying the role of the board for current and new members of the board, such a statement also clarifies the role of the board for the medical staff and management, and helps to outline the relative roles and responsibilities between those groups.

The defined statement of the board's role in quality and QA will also have a definite and useful purpose regarding the reporting of quality and QA information to the board. Such a statement of the board's role will provide the framework for what information is communicated to the board and how the board can appropriately respond to that information. This is a critical step in insuring that the board actually oversees and directs the QA program and quality of care of the hospital by defining its role and by defining the information it needs to support and fulfill that role.

Although different boards will have different statements of roles regarding quality and QA, the statements should have common characteristics. These include: a statement of the board's responsibility for quality; a statement of the hospital's commitment to quality and the board's role in assuring that commitment; a statement of the board's responsibility for overseeing the activities of the medical staff, with specific attention to its role in medical staff credentialing; a statement of the board's level of involvement in the hospital's QA program and quality-related activities; and, reference to *how* the board will discharge its role in quality and QA.

Again, it is important to stress that the role statement should be specific to each individual hospital and board. The statement will depend to some degree on the current level of board involvement in quality and QA as determined by the assessment instrument presented in chapter 5.

Following is a *sample* statement of a board's role in quality:

"The XXX hospital is committed to providing quality care to all of its patients and it is the governing board's responsibility to assure that the hospital fulfills that commitment. The board has the ultimate responsibility for the quality of care provided by the hospital; for the hospital's quality assurance and risk management programs and all quality improvement activities; and, for the oversight of the hospital's medical staff, which is primarily discharged through its oversight of and final decisions regarding the appointments and privilege delineations of medical staff members.

"The board will oversee and direct the hospital's quality assurance program and related functions by regularly monitoring specified information regarding the effectiveness of the QA program and related functions. This information will place specific emphasis upon the QA program's effectiveness in the identification and resolution of problems in care, in the improvement in the quality of care, and in the effectiveness of the participation of the medical staff in QA activities. The board will monitor to assure that the results of clinical QA reviews are well and objectively integrated into the medical staff credentialing process. The board will monitor the hospital's quality of care primarily through its regular review of a number of defined quality indicators, as well as other information that may be appropriate.

"Overall, it is the role of the board to monitor the QA and credentialing processes of the hospital to assure that they are functioning effectively. The board will take action when it determines that these processes are not functioning effectively, or that they are in need of improvement. Similarly, it is the role of the board to monitor the quality of care, and to take action when there are negative trends in that care or when the quality of care is in need of improvement."

DEVELOP GUIDELINES FOR THE TYPE OF INFORMATION TO GO TO THE BOARD

Once the role statement is complete, the next step is to develop guidelines for the content and formats of QA and quality-related information that will go to the board. These guidelines should address QA program and process information; quality-related program and process information (such as risk management, infection control, utilization review, etc.); information about the quality of patient care (this will relate to the next step, the development of quality indicators for the board); information about external sources of quality-related information; reporting formats; and reporting frequency to the board.

Again, guidelines should be specific to each hospital and board, and should be used to structure the information that will flow to the board.

Sample Guidelines for Reporting of Quality Information to the Board

1. The board will receive information on the effectiveness and completeness of the QA program and its processes at least twice per year.

2. The board will receive information on the effectiveness and completeness of department-specific QA processes periodically, with attention devoted to those departments where the QA processes are not functioning completely or appropriately. Emphasis will be placed on medical department review functions.

3. The board will receive information on the quality of patient care at every board meeting.

4. Information on the quality of patient care will be provided to the board in summary fashion that compares data over time and that tracks trends in graphic formats whenever possible. Information presented will be compared to preestablished threshold or target levels.

5. The quality of care information presented to the board will be rotated from meeting to meeting consistent with the preestablished board quality calendar.

6. Specific quality problems will be presented to the board in the following situations: to demonstrate the QA problem resolution process of a specific department or function; when a problem that poses risks to patients has gone unresolved for a period of three months; or, when a problem that poses significant risk of injury or death to patients, families, or staff, or liability to the hospital is identified.

7. Unless otherwise requested by the board, minutes of medical

staff or departmental meetings will only be presented to the board as background for other quality-related reports.

8. The board will be presented with detailed information to facilitate its appraisal of the QA program over the course of two consecutive board meetings once every year; the risk management program will be assessed during this same process and relevant information provided.

9. The board will be presented with all JCAHO survey reports, with all Type I recommendations (contingencies) highlighted.

10. The board will be presented with all PRO quality sanctions, Health Care Financing Administration Medicare Mortality reports, department of health survey reports, liability insurance surveys, and reports of pending and settled malpractice claims.

11. The board may request specific QA or quality-related information at any time.

12. The board will be presented with Medical Executive Committee (or equivalent) recommendations for appointments and reappointments to the medical staff and delineation of clinical privileges in the following manner: for each applicant there will be a summary sheet that contains the MEC, credentials committee, and department head recommendation compared to the verified compliance or noncompliance with the explicit criteria for medical staff membership and privilege delineation.

13. The board will be presented with a manageable number of medical staff credentialing cases each meeting, not to exceed twenty-five cases per meeting.

The guidelines presented for quality reporting to the board are presented as sample guidelines only. They should *not* be implemented for any board without serious review, evaluation of appropriateness, and modification and refinement. The guidelines should be tailored for each board and will depend upon many factors, not the least of which is the board's current level of involvement in quality. The guidelines will evolve over time, and will be refined and modified to reflect the board's evolving commitment to and involvement in quality.

The value of such guidelines is in structuring the information that flows to the board. By doing so, the role statement of the board is made operational and understandable. Further, arguments or disagreements concerning what information should go to the board will be addressed *prior to* the occurrence of specific quality problems, and thus will be depersonalized and more easily dealt with.

DEVELOP A SERIES OF QUALITY INDICATORS FOR THE BOARD

The next step in the process is to develop and refine a series of quality indicators that are appropriate for the board and that support the board's defined role in quality and QA. The indicators will provide the board with information about specific quality issues or areas in the hospital that have been identified as important and worth reviewing. Specific indicators chosen should reflect the perspectives that frame the hospital's definition of quality as discussed in chapter 4. Although many of the indicators will reflect clinical issues, they *should not* be solely limited to those and should include other issues relating to quality as well. The indicators should be based primarily on internal data or information which is valid and reliable and should be presented in formats that demonstrate trends over time.

Much of the raw material for quality indicators for the board will be quite familiar to QA professionals and will come from sources such as generic screening or occurrence screening criteria, incident report criteria, malpractice liability insurer mandated reportable events.

Quality indicators for the board can be developed to provide the board with the overall picture of quality for the whole hospital. Additionally, quality indicators can be developed and tracked occasionally for specific departments or areas within the hospital.

The board quality indicators presented by no means represent an exhaustive or exclusive listing of all the potential quality indicators appropriate for a hospital board. They are presented as examples of the various types of possible quality indicators that can be used by a board.

A valuable process for determining which initial quality indicators should be selected for routine board monitoring includes developing an extensive list such as that presented. Next, each of the possible indicators on the list should be reviewed with the board or the board QA committee or equivalent. The potential indicators should be reviewed for meaning to the board, relevance to the hospital's approach to quality and QA, and availability of valid and reliable data.

Next, the list of indicators should be narrowed down to a list of no fewer than twelve and no more than twenty quality indicators. The final list should be reviewed again to insure that it does not reflect only the clinical perspective of quality, and that all of the indicators together form a reasonable overall picture of the hospital's quality of care from the perspective of the board.

Remember that the indicators will not be the only information the board receives about quality, they will simply be the governance information about quality that the board routinely monitors.

Although it is important to choose the indicators with care, they will not be cast in stone. In the future, new indicators can be added, and original ones can be deleted or modified. This refinement process should be based on the value of the information to the board and how indicative of quality each indicator truly is. Over time, it will become apparent that some indicators are more meaningful than others.

It may become evident that indicators that were useful in the past are no longer as valuable because the levels of quality have been improved and are being maintained at optimal levels. In this situation, the board may wish to turn its

Examples of Quality Indicators for the Board

- Hospital-wide mortalities (hospital wide, by department or area)
- Nosocomial infections (hospital wide, by department or area)
- Post-operative infections
- Patient complaints (hospital wide, by department or area)
- Patient satisfaction survey report trends
- Staff turnover and/or absenteeism presented by department or area
- Staff complaints
- Staff satisfaction survey report trends
- Medical staff complaints
- Medical staff satisfaction survey report trends
- Patient falls
- Adverse drug reactions or interactions
- Unplanned returns to surgery
- Unplanned transfers to surgery, isolation, intensive care units, or cardiac care units
- Unplanned transfers to other acute care facilities
- Hospital-incurred traumas
- Discharges against medical advise (AMAs or "elopements")
- Returns to the emergency room within 72 hours of being treated in the ER
- Readmissions to the hospital within one month of discharge
- Unplanned admissions to hospital following outpatient procedures
- Assaults or altercations (patient to staff, patient to patient, staff to patient, staff to staff)
- Cesarean section rates
- Neonatal and maternal mortalities
- Perioperative mortalities
- Medication errors
- Patient wait times by department or area
- Denials of payment
- Content analysis summaries (i.e., positive, negative, or neutral ratings) of newspaper and electronic media stories about or relating to the hospital
- Pattern analysis of incident reports or occurrence screens
- QA process measures

attention to other quality indicators that demonstrate areas needing improvement, and occasionally monitor the former indicators to insure that quality is being maintained at acceptably high or optimum levels.

The actual process of choosing and continually refining quality indicators will be of great importance in enhancing the board's understanding of and commitment to quality, and in building the relationship between the board and the QA professional.

Determine Threshold Levels for the Quality Indicators

Once the quality indicators for the board have been selected, the next step is to choose threshold or target levels for each quality indicator. This is a critical step in making the indicators meaningful measures of different aspects of quality, and in taking the indicators beyond being simply information to being *tools for improving quality*.

Presenting a specific quality indicator over time to a board will be meaningful to that board in terms of demonstrating whether the specific quality indicator is getting better or worse over time. Comparing the indicator to a pre-determined threshold or target level, however, will instantly tell the board two key things. First, it will demonstrate whether the quality of care covered by the specific indicator is *currently acceptable*. Second, if the quality indicator exceeds the threshold or target level, it will tell the board that *action must be taken* to address the issue and improve the quality of care.

More specifically, an exceeded threshold or target level will demand *board action*. Hence the term "threshold level," as it will become a threshold for action by the board. In fact, calling the target levels "thresholds for action" is useful in keying the board in to its responsibility to respond to negative trends in quality indicators. When a threshold is exceeded for a specified length of time, the board should act. This action could be limited to the board asking why and seeking more information, or it could involve the allocation of resources to the area in question, or more direct action such as enforcement or revision of policies and procedures, or disciplinary action, or any number of responses appropriate to the situation.

The essence of this process is transforming the quality indicators into *tools for continuously improving quality.* The threshold levels do this if they are not considered static once set, but are continually tightened to reflect improvements in trends in quality. Thus, appropriately tightened threshold levels may elicit higher and higher levels of quality.

For example, if hospital wide mortalities have been well below the threshold level for the past two quarters, the board could *lower (tighten) the threshold level.* In so doing, the board will be setting a higher standard and will transmit this message through the organization. The board could then direct the hospital's quality resources to investigate methods of process refinements that could reduce mortalities. By doing this the board will be applying statistical process control measures, will be sending a message of strong commitment to quality throughout the hospital, and will be actually taking governance-appropriate steps to reduce hospital wide mortalities to minimum levels.

Of course, it is possible to set threshold levels *too tight,* and thus they might be occasionally loosened (say the mortality threshold raised from 2.5 to 3.5 percent). This only must be done when it is clearly evident that the threshold levels were set too tightly. *It must never be done to accommodate and justify consistently poor levels of quality as demonstrated by a specific quality indicator.* This important point clearly raises the question of *how* threshold levels are determined.

Threshold levels can be determined in a number of acceptable ways.

- published national averages (for example, an average hospital wide nosocomial infection rate published by the Centers for Disease Control or some other national, reputable organization; incident report ratios [number of incidents per bed per day] from malpractice insurers)

- accepted and published clinical norms or averages published by professional or national organizations (for example, the average C-section rate)

- the average for the quality indicator for the specific hospital over the past several years; or a standard deviation from the mean

- *goals* for quality improvement, (for example, reduce C-section rate to 18 percent, or reduce nosocomial infection rate to 1 percent) (Quality Letter for Healthcare Leaders, September 1989)

- averages from similar hospitals (for example, university-affiliated teaching hospitals, small-rural hospitals, hospitals within a multi-hospital system, etc.)

It is critical to again stress that once determined, the threshold levels *will not be static*. To truly be quality improvement tools, the threshold levels should be adjusted periodically to reflect higher quality indicator goals and to reflect improvements in quality indicator trends.

The threshold levels will be even more meaningful to the board if the board, or appropriate board committee—such as the board QA committee, participates in setting the threshold levels for each quality indicator. This interactive discussion with the QA professional will reinforce to the trustees involved the value of the quality indicator and will clarify and emphasize the meaning of the threshold level. Further, the board should always be involved in the revision of established threshold levels.

The issue of threshold levels for the quality indicators is addressed further in the next section on reporting formats.

Report Formats for the Quality Indicators

As has been alluded to throughout this and the earlier chapters, there are several relatively simple guidelines for reporting quality information to the board which the reporting formats should adhere to and support.

Quality report formats to the board should:

- include predetermined board quality indicators
- be in graphic form whenever possible
- demonstrate trends over time
- reflect predetermined threshold or target levels for the quality indicators whenever possible
- contain a minimum of narrative
- be meaningful to the board, i.e., contain *governance information*
- focus on outcomes
- when presenting problems, do so using the five-step problem focused process to facilitate board understanding.

The following figures present examples of report formats for quality indicators that incorporate the preceding points.

Figure 6.1 shows a graph that demonstrates hospital-wide mortalities over a two year period. Note that by using the graphic format, trends and peaks or valleys are immediately discernable. Thus the board can meaningfully focus in and ask appropriate questions about the information. Figure 6.2 shows the same graph with the addition of a threshold line. Notice the visual power of the threshold line, as it transforms the graph into better governance information for the board.

134 QUALITY FROM THE TOP

**HOSPITALWIDE DEATHS
as a percentage of discharges**

Figure 6.1

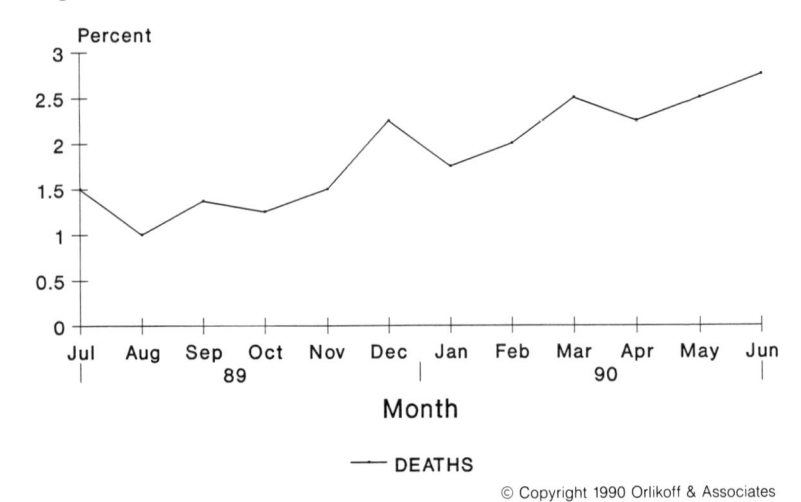

© Copyright 1990 Orlikoff & Associates

Figure 6.2

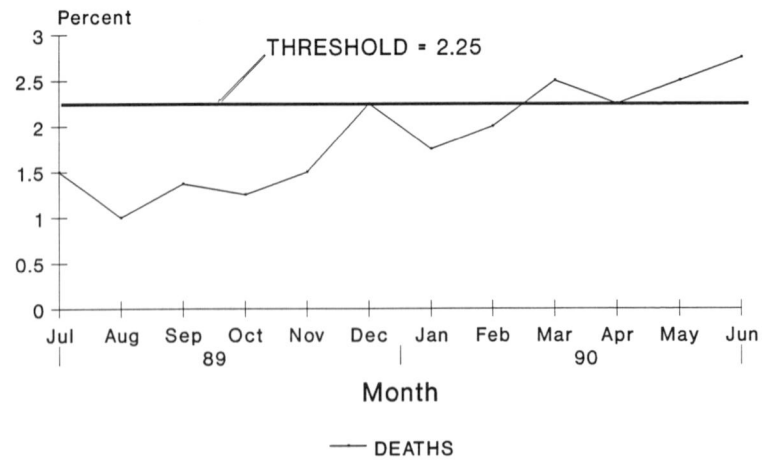

THRESHOLD = 2.25

© Copyright 1990 Orlikoff & Associates

Meaningful Quality-Related Information 135

Figure 6.3 shows a graph of nosocomial infections. In figure 6.4 a threshold line of a 4 percent nosocomial infection rate is added. Clearly, the threshold line in Figure 6.4 is much too high (too loose) and the illusion is perpetrated that the nosocomial infection rate is acceptable and well below target levels. Figure 6.5 is the same graph with a different threshold line that is set lower at 2.5 percent. This is a much more appropriate threshold line, and will draw the board's attention rather than its complacency.

Figure 6.3

**HOSPITALWIDE NOSOCOMIAL INFECTIONS
as a percent of discharges**

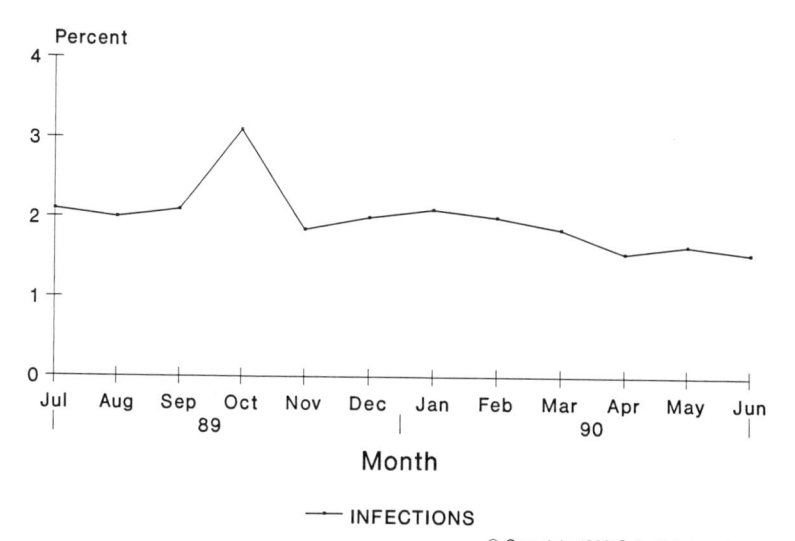

© Copyright 1990 Orlikoff & Associates

HOSPITALWIDE NOSOCOMIAL INFECTIONS
as a percent of discharges

Figure 6.4

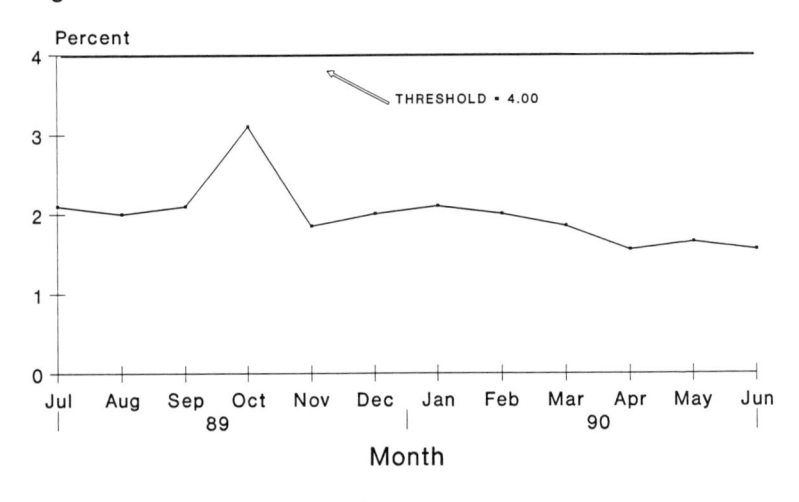

THRESHOLD = 4.00

INFECTIONS

© Copyright 1990 Orlikoff & Associates

Figure 6.5

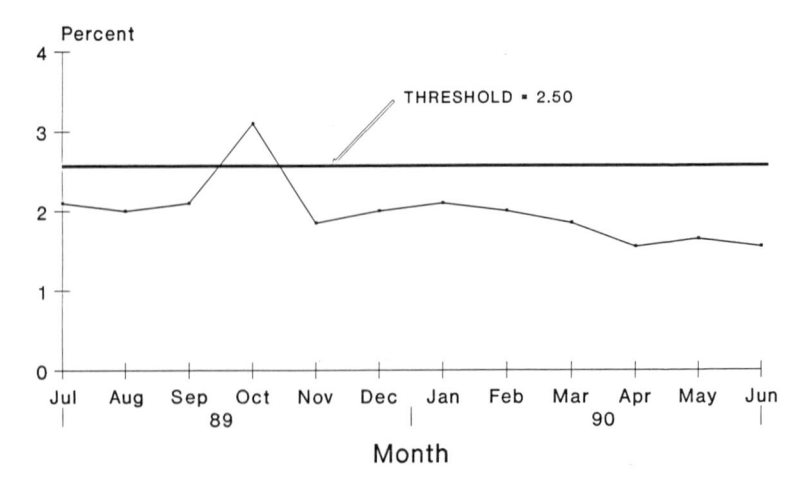

THRESHOLD = 2.50

INFECTIONS

© Copyright 1990 Orlikoff & Associates

Consider this situation: A board, by monitoring hospital-wide mortalities as a quality indicator, noticed an increase in the overall hospital mortality rate and asked the QA professional to locate any departments where mortalities had been increasing significantly. Once the department of surgery was identified, the QA professional provided the board with a graph of mortality rates for surgery, and with the board determined a threshold line for the focused quality indicator of surgical mortalities. That graph and threshold line is presented in figure 6.6.

Figure 6.6 demonstrates the profile of a threshold that may be too low, or too tight. In other words, the threshold line in figure 6.6 may present a goal or target that is unattainable and therefore inappropriate. Figure 6.7 shows that same graph with a slightly relaxed, and probably more realistic threshold line.

Figure 6.6

**DEPARTMENT OF SURGERY MORTALITY RATES
as a percent of surgical cases**

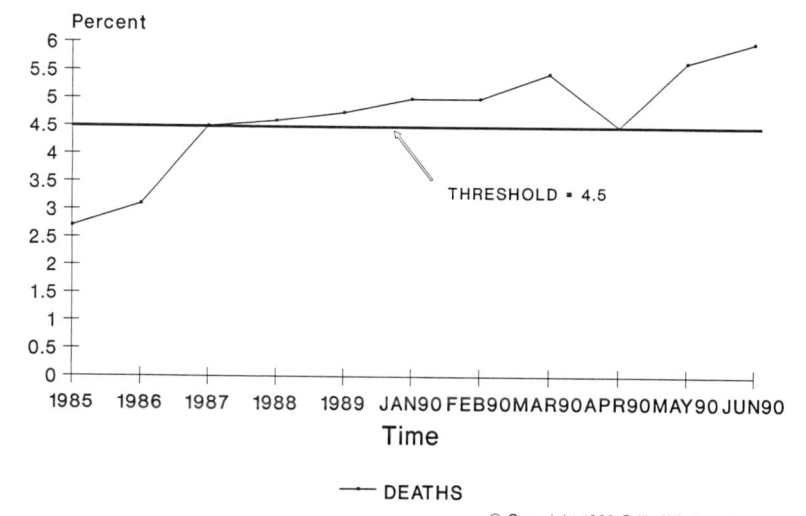

THRESHOLD = 4.5

— DEATHS

© Copyright 1990 Orlikoff & Associates

Figure 6.7

**DEPARTMENT OF SURGERY MORTALITY RATES
as a percent of surgical cases**

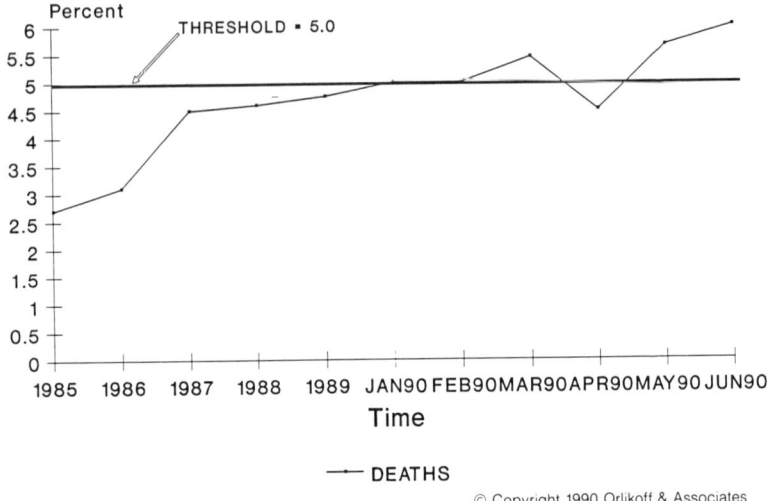

[Graph showing mortality rates with THRESHOLD = 5.0, plotting DEATHS from 1985 through JUN90]

© Copyright 1990 Orlikoff & Associates

Clearly, the setting of the threshold levels is critical. It is important to remember, however, that the thresholds have meaning in relation to the *trends* of the quality indicators. When certain quality indicators are graphed, the profile of the graph (such as one with a severe spike, or bi-polar distribution) will be such that a meaningful threshold will be difficult or impossible to create.

Figure 6.8 shows a graph of the average patient waiting time for X-ray. This graph shows that there are two half-hour periods during the day when patients scheduled for X-ray have wait times of seventy minutes or more. But the graph also shows that during the rest of the day the average wait time is always twenty-five minutes or less. Where would a meaningful threshold line be placed on this graph?

Figure 6.8

AVERAGE PATIENT WAIT TIME FOR X-RAY

Wait Time

© Copyright 1990 Orlikoff & Associates

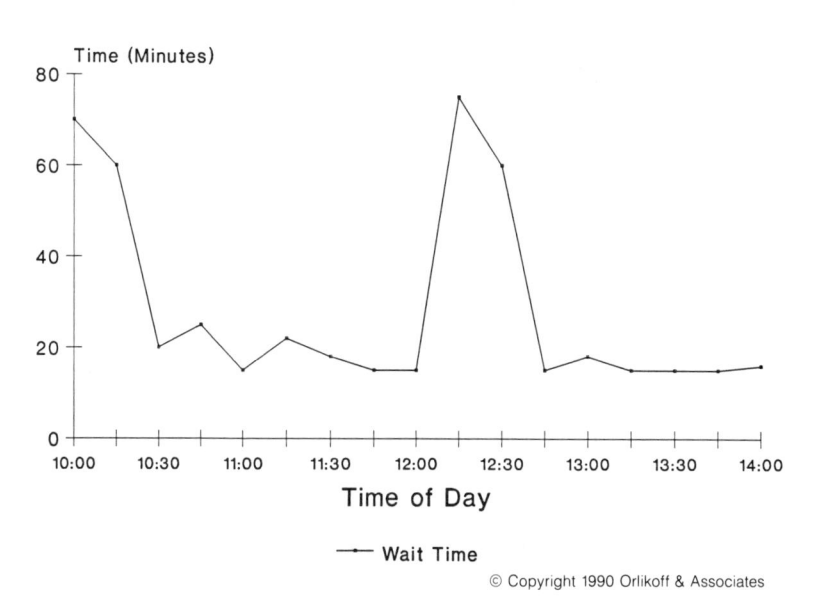

Using the arithmetic mean would place the threshold line at about thirty-three minutes, well above the vast majority of waiting times, but well below the two seventy minute waiting times.

In this situation, using the arithmetic mean as the threshold line may not be best as it really would not provide a meaningful reference point. Here, once the process affecting waiting times was thoroughly investigated, a better threshold line might be a *goal* for waiting times, which might be below the average wait time.

The preceding figures demonstrate the power and effectiveness of providing boards with quality indicators in graphic formats with threshold lines. They are classic examples of *governance information.* Compare them against any set of minutes, or most narrative reports and then ask which format provides the best information for boards.

A good test of the effectiveness of the format of quality reports to boards is that they should "RUMBA." That is, they should be: Realistic, Understandable, Measurable, Behavioral (i.e., outcome focused), and Achievable (i.e., present realistic, achievable thresholds). If the quality report formats pass the "RUMBA" test, as graphed quality indicators with threshold levels do, then they are likely to be appropriate and meaningful to the board.

Define the Other Quality-Related Information to Go to the Board

Even though the quality indicators and threshold levels will probably be the most useful quality information to go to the board, it is important to remember that other QA and quality-related information will also need to be transmitted to the board. These other types of information should also be formatted in such a way as to insure that they provide meaningful governance information.

The next step is to define what other (non-quality indicator) QA and quality-related information will go to the board. Again, this discussion should be conducted with the QA professional and the board or the appropriate board committee. In this way, the trustees involved will get an understanding of what information is available about quality and QA, which information will be most meaningful to the board, why the board should receive certain information, and why the board must actively respond to that information. Furthermore, the QA professional will gain an understanding from this discussion about the board's current level of awareness and understanding of quality and QA.

It is important to remember, as was pointed out in chapter 4, that the board is also responsible for the effectiveness of the QA program. Thus, some of the other information presented to the board should relate to the effectiveness of the QA program and its various processes.

Certain types of QA and quality-related information, other than quality indicators, *must* go to the board. These types of information include:

- Medical staff appointments, re-appointments, privilege delineations, privilege restrictions or revocations
- Information at least every six months on the frequency, severity, and causes of adverse patient occurrences (JCAHO, 1988)
- Information at least every six months on the actions taken and their results to reduce the frequency and severity of the adverse patient occurrences (JCAHO, 1988)
- Information relating to the annual reappraisal of the QA program, including the QA plan.

In addition to the information that must go the board, there is quite a bit of other quality and QA information, both within the hospital and external to the hospital, that will be periodically useful to the board. Chapter 4 provides a detailed sample of the types of non-quality indicator information about quality and the QA program that can go to the board.

This information, in addition to coming from the QA program, should also be extracted from the risk management, infection control, patient representative, and other internal programs.

In addition, external sources of information about quality and QA should be considered for occasional or routine presentation to the board. Such external information includes:

- Reports of JCAHO accreditation surveys
- Medicare mortality data released by the Health Care Financing Administration

- Professional Review Organization reports, including sanctions
- Malpractice insurer surveys and reports
- National Practitioner Data Bank information (integrated into medical staff credentialing information)
- State licensure survey reports

As the rather extensive universe of other quality-related and QA information that *could* go to the board is defined, it will become clear that not all of it *should* go to the board. Some information may be routinely sent to the board while other information will only occasionally or infrequently be transmitted to the board.

Again, it is important to determine why certain information should go to the board and then to develop the best format to send the information to the board. The choice of the other quality-related and QA information that goes to the board will vary tremendously from hospital to hospital. It will be a function of the results of the assessment of the board involvement in quality (chapter 5), of the current state of the hospital's QA program and quality of care, and of the specific problems or issues relating to quality and QA that will be different from hospital to hospital.

In choosing these types of information and presentation formats for the board, it is critical that they meet the criteria outlined earlier in this chapter for effective quality reports to the board, and that they *do not* succumb to the common weaknesses of quality reports.

Information should never be sent to the board simply because its there and can be sent. The other quality-related and QA information to the board must be organized around the defined role of the board regarding quality and QA. The information must be measured against the guidelines for quality reporting to the board that will be developed by each hospital.

Develop a Calendar for Quality Reporting to the Board

As should now be clear, the volume of quality-related information that may find its way to the board can easily be rather substantial, if not overwhelming. To avoid burying the board in quality reports, and, more importantly, to integrate all the quality-related information to the board, it is useful to develop an annual calendar of quality and QA information to the board. This calendar will simply outline an annual schedule of the quality and QA information that will go to the board at each board meeting throughout the year.

The calendar should reflect both the board's role statement regarding quality and QA, annual goals relating to quality and QA, and should practically express the components of the board's annual workplan relating to quality and QA. Further, if there is a board QA committee or equivalent, the calendar should be linked to the board QA committee workplan and calendar.

One purpose of the calendar is to spread out the information the board will receive and to prevent the board from becoming inured from seeing the same information in the same format over and over at each board meeting. Another purpose is to make certain that the board gets an integrated, oversight picture of the hospital's quality of care and QA program.

To construct the calendar, certain questions must be asked and answered. For example: Is the medical staff credentialing process flexible enough to allow for the recommendations to the board to be divided into manageable segments throughout the year? How often should the board see the quality indicators? Should all the quality indicators be presented together, to facilitate an integrated picture of quality for the board, or is there some reason that they should be presented independently of one another and at different times? Are there certain quality-related issues or areas that the board should focus on more than others, and if so should these be presented to the board more frequently?

Once these and other related questions have been answered, the board quality calendar can be developed and implemented.

Sample Board Quality Calendar

JANUARY: Approve annual board quality and QA workplan; affirm quality indicators and threshold levels; review quality indicators; medical staff credentialing.

FEBRUARY: Semi-annual report on risk management program, including information on: frequency and severity of adverse patient occurrences, actions taken to minimize future occurrences, results of actions, review of pattern analyses of incident reports compared to malpractice claims, review of open and closed malpractice claims; medical staff credentialing.

MARCH: Review of Patient Representative program; review of medical staff departmental QA functions; medical staff credentialing. Board retreat to focus partially on quality and QA.

APRIL: Review quality indicators; medical staff credentialing.

MAY: Review of infection control program; focused review of least effective medical staff department QA function; medical staff credentialing. Continuing education program on quality and QA.

JUNE: Review of nursing QA program and functions; review of recent external quality-related reports (PRO, state, multi-hospital system surveys, etc.); medical staff credentialing.

JULY: Review quality indicators; medical staff credentialing.

AUGUST: Review semi-annual risk management report; medical staff credentialing.

SEPTEMBER: Review ancillary department QA functions; staff attitude surveys, staff turnover reports, staff complaints; medical staff credentialing. Continuing education program on quality and QA.

OCTOBER: Review quality indicators; review JCAHO survey preparation findings; medical staff survey results; medical staff credentialing.

NOVEMBER: Overview of QA program as Part I of annual QA program reappraisal; review this years Medicare Mortality data in relation to past years; medical staff credentialing.

DECEMBER: Complete annual QA program reappraisal including review and approval of revision of QA plan; review draft board quality workplan for next year; begin review and evaluation of quality indicators and threshold levels.

This sample calendar demonstrates that the fictitious board will receive all the board quality indicators together, every four months. It further shows that the board will consider medical staff credentialing decisions at eleven of its twelve meetings. It then reflects other quality and QA related information that will be presented in rotation throughout the year to the board, as well as continuing education for the board about quality, QA and related topics.

Please keep in mind that the annual board quality calendar presented here is a sample calendar only. Each board will have a very different quality calendar based upon many different and individual characteristics. Further, the more the board quality calendar reflects the individual goals, sophistication in quality and QA, and specific characteristics of the board, the more meaningful the calendar will be. Each calendar must be specifically tailored by the QA professional to their own board, and should flow directly from the preceding steps outlined in this chapter.

Final Steps

The final steps in the process are fairly self-explanatory. Rather than being discrete, they are integral to the whole process. The final steps in the process are to:

- Monitor and evaluate the effectiveness of the quality indicators, and refine them;

- Monitor and evaluate the appropriateness of the quality indicator thresholds, and modify the thresholds as appropriate; and,

- Evaluate the entire process and refine as appropriate.

The first two points were discussed earlier in the sections of this chapter devoted to quality indicators and threshold levels. It is worth stressing, however, that the quality indicators and thresholds, and the very process itself, should not be static. To be effective, the process must be a dynamic one; one that is continually evaluated and modified and improved.

That leads to the final step, the evaluation of the process and the implementation of appropriate refinements. Although the evaluation and modifications should be ongoing, there is value in building in a discrete evaluation module into the process. This will help formalize the evaluation of the process and help prevent the process itself from becoming immutable and stale.

Thus, it is worth the effort to conduct a fairly formal evaluation of the process annually. This evaluation can most logically be incorporated into the annual QA Program reappraisal. The evaluation itself need not be extensive or exhaustive. It should simply assess the board's perception of the strengths and weaknesses of the process, as well as the perceptions of the QA professional. It is also possible to fold this evaluation into a reassessment of the board's role in quality and QA (chapter 5), although it is unlikely that the detailed assessment of the board's role in quality will be conducted annually.

The essence of the evaluation is the improvement of the process: the improvement of the quality indicators, of the thresholds, of the reporting formats, of the quality calendar, of the information to the board. As the information to the board regarding quality and QA is continually refined and made more meaningful as governance information, the board will more effectively oversee the quality of care and QA program.

CONCLUSION

The involvement of any board in quality and QA will only be as good as the information that is presented to the board and the way it is presented. Good governance information presented in effective and understandable formats will *elicit* good board involvement in and oversight of quality. Equally important, good governance information regarding quality will build and maintain strong board commitment to quality and QA. Furthermore, it will build a strong and productive partnership between the board and the QA professional.

For these reasons, it behooves the QA professional to take a long and structured look at the information that is currently being presented to the board regarding quality. By systematically improving the content and format of that information, the QA professional will systematically improve board involvement in quality, and will empower the board to actually oversee and direct and improve the hospital's quality of care.

References

1. Passett BA. "Boards should set and monitor standards of quality." The Quality Letter for Healthcare Leaders. 1:1 Jul–Aug 1989. Bader & Associates.
2. "Board quality indicator report." The Quality Letter for Healthcare Leaders. 1:2 Sep 1989. Bader & Associates.
3. "Scoring guidelines for hospital risk management activities." Joint Commission on Accreditation of Healthcare Organizations. Chicago, 1989.

CHAPTER SEVEN

Quality and Board Structure

INTRODUCTION

Board structure, or organization, relates to such things as committees of the board, size of the committees, reporting relationships, and so on. Structure is important in that it *supports* function. That is, the structural or organizational components should facilitate the maximum level of function of the organization in question.

Unfortunately, as logical as that may seem, there are many who believe that structure actually *creates* or drives effective function. An example of this thinking can be seen in the common hospital response to the need to develop better board oversight of quality: "establish a board QA committee." Simply altering existing structure by creating a new committee will not by itself improve function. In fact, structural change can either facilitate function, or it can inhibit function. Clearly, the goal of structural change should be to facilitate effective function.

Once the QA professional and the board have determined what the proper function and role of the board regarding quality is, and what information will flow to the board to enable it to perform that function, then they can consider what board structure will best support and facilitate that function.

STRUCTURAL OPTIONS FOR BOARD INVOLVEMENT IN QUALITY

As alluded to earlier, the most common response to the question of improving board involvement in quality is to establish a board QA committee or its equivalent. While this may be appropriate, most hospitals do this without first defining the role of the whole board in quality, the role of the board QA committee and its relationship to the board and to the hospital QA program and related functions. This common tendency rarely results in improved function as it is generally structural change for its own sake.

There are several structural options. They include:

- A board quality committee
- Delegating the function that the board quality committee would perform to an existing committee, such as the Joint Conference Committee
- Having no committee of the board performing quality functions and having the board perform the functions entirely
- Having no board committee performing quality functions, having the whole board perform the quality functions, but having trustees assigned as members of standing hospital or medical staff committees that deal with quality, such as the hospital QA Committee and the Credentials committee
- Any conceivable combination of the above points.

Each QA professional and board must consider their own individual characteristics and circumstances to determine what structural option is most appropriate. There is no reason to change the current structure just for the sake of changing it. The only reasons to implement structural changes are if the current structure is inhibiting effective board involvement in quality, or if a different structure will significantly improve that function.

Sometimes there are compelling reasons to have a board QA committee, such as a board that is required by law to have meetings open to the public and press. Here, the board might not be able to consider sensitive quality matters effectively without the protection and confidentiality that the structure of a board committee may provide. Another often cited reason for a board QA committee is a large hospital that generates enormous amounts of quality information where the board might be buried in quality issues unless a committee did some of the background screening and legwork.

The important thing to consider is *why* creating a board QA committee would *facilitate* the board's involvement in and oversight of quality and QA, and then to determine *how* the committee will do that. In fact, those questions should be asked and answered prior to the implementation of any structural changes for the purpose of strengthening the board's quality function.

A Process for Creating an Effective Board QA Committee

The answer to the question of why a board committee would be beneficial becomes the charge or stated role of that committee.

Next, the function of the committee in relation to that of the board should be explicitly defined, as well as in relation to the hospital's QA program and medical staff and such activities as medical staff credentialing.

The next step is to define the composition of the committee. A strength of a board committee is that it can have members who are not members of the board to broaden its perspective and bring it useful expertise. Any board QA committee should have members who are drawn from the following groups: the board, the medical staff, top management, nursing, and the QA professional. Often, it will be the QA professional who will staff this committee. The QA professional should also have significant input to the committee, and staffing the committee may facilitate this. Unfortunately, the demands of staffing the committee will often preclude effective QA professional involvement. Here, the QA professional should be made a member of the committee with other provisions made for committee staff functions.

A final point regarding committee composition relates to the number of board members on the committee. Even though it is wise to have several non-board members on any board QA committee, it is important to remember that it is a committee of the board and thus should be composed of a majority of board members. This will involve as many trustees as possible, and will prevent trustees from being intimidated or dominated by other members of the committee.

Once the charge is defined and the composition of the committee established, the next step is to define the specific responsibilities of the committee. As discussed in earlier chapters, this will relate to the content and format of information sent to the committee, and what information the committee will send to the board and other groups within the hospital. The issue of the authority of the committee should be addressed at this time. Generally, a board QA committee will have no authority as such, other than making recommendations to the board. Regardless of what structures are chosen, the board always retains the ultimate authority and decision making mandate.

Often, board QA committees are charged with the responsibility of screening QA information before it goes to the board. If this only results in the board seeing the same information that the committee has seen then it is a waste of time. The committee must have a specific set of responsibilities if it is to facilitate the board's involvement in quality. Of course, the specific responsibilities of any board QA committee will flow from and support the explicitly defined role of the board regarding quality as discussed in chapter 6.

Conclusion

This chapter has purposely not outlined how to establish the various structural models relating to increasing a board's involvement in quality. This should flow from the defined role and function of the board regarding quality and related issues. Once the function is defined and made operational through information content, format, and flow, then the structure that best supports this function can be considered. In fact, once the role and function and information requirements have been addressed, the appropriate structure to facilitate effective function will often be self evident.

CHAPTER EIGHT

The Future of QA Professionals

The 1990s will be a period of challenge and stress for hospitals and their boards. The cost pressures faced by hospitals in the 1980s are now evolving into severe pressures to provide quality within diminishing resources. Hospitals in competitive environments will find that they are competing on quality as well as on price. At the same time, external regulation relating to quality will increase and put hospitals under still more pressure.

A hospital's ability to deliver quality of care and to compete on quality will be a function of its *internal* commitment to quality, and *not* external quality regulation. The hospital's true internal commitment to quality will directly result from and be reflected by the level of its board's commitment to quality. It will also be practically reflected in the level of board involvement in and oversight of the hospital's quality of care, QA program, and medical staff credentialing activities.

This commitment on the part of a board will rarely develop by itself. It must be nurtured through a period of education and evolution, and through a systematic process of making quality understandable and meaningful to the board. The board must realize through experience that it *can* have a significant impact upon the quality of care provided by its hospital, and that it *must* be meaningfully involved in quality if it is to do so.

Similarly, QA professionals must find avenues to rise above the limitations of most current hospital QA programs to become true leaders and change agents in their hospital's quest for quality. The best avenue for achieving this is for the QA professional to first become an educator to their board about quality and QA, then a consultant to their board, and finally a partner of the board in the quest for quality.

The QA professional can and should be the one to guide the board's evolution to an effective quality consciousness. The QA professional can facilitate their board's understanding of its responsibility for and role in quality. The QA professional can practically help their board to effectively define and perform its role in quality and to accept and discharge its responsibility for quality.

In this way the QA professional can become an effective instrument in creating and maintaining a sincere organizational commitment to quality. In addition to improved quality of care, this will also result in improved organizational visibility, support, and authority for the QA professional.

This book has attempted to demonstrate how to assess and systematically improve a hospital board's commitment to and involvement in quality. True change in an organization's culture or in its commitment to quality must come from the top of the organization and must continually be supported from the top of the organization.

Some QA professionals will find the task of improving their board's involvement in quality easier than others. Yet even for those QA professionals who face an unreceptive board and management the task, while difficult, is not impossible. With analysis, planning, effort and patience the QA professional can enhance their board's commitment to and involvement in quality. In so doing, their own position, performance, and satisfaction will also be enhanced.

Quality is often described as a journey, not a destination. When quality is a valued priority of the hospital, the journey will be an easier and more productive one. Furthermore, the QA professional will be a leader in the journey, and a valued and recognized asset to the board and the hospital.

APPENDIX A

Leadership Responsibilities and Functions of the Hospital Governing Board

THE AMERICAN HOSPITAL ASSOCIATION, as do most state legislatures and regulatory agencies, portrays the hospital board as the ultimate authority of the hospital which has the final responsibility for patient care and quality. The board has the leadership responsibility for the hospital but must, through negotiation and delegation, share this leadership responsibility with executive management and the medical staff. The AHA (1982) defines six broad categories of the board's leadership responsibilities and functions:

1. *Organization.*

    The Governing board has the responsibility for organizing itself effectively, for establishing and following the policies and procedures necessary to discharge its responsibilities, and for adopting bylaws in accordance with legal requirements.

    The governing board has the responsibility for selecting a qualified chief executive officer and for delegating to the chief executive officer the necessary authority to manage the institution effectively.

The governing board has the authority and responsibility for ensuring proper organization of the institution's medical staff and for monitoring the quality of care provided under the auspices of the institution.

2. *Public Policy and External Relationships.*

   The governing board has the authority and responsibility for monitoring and influencing public policies concerning the delivery of health care and for ensuring the establishment and maintenance of appropriate external relationships.

3. *Strategic Planning.*

   The governing board has responsibility and authority, subject to the institution's charter, for determining the institution's mission and for establishing a strategic plan, goals, objectives, and policies to achieve that mission.

4. *Resource Management.*

   The governing board is entrusted with the resources of the institution and with the proper development, utilization, and maintenance of those resources.

5. *Human Resources Development.*

   The governing board has the responsibility and authority for the organization, protection, and enhancement of the institution's human resources.

6. *Education and Research.*

   The governing board is responsible for the provision of health care educational and research programs that further the hospital's mission.

The AHA's 1982 overview of the roles and functions of the hospital board presents a reasonable and broad overview of the board's general responsibilities. Notice, though, that the board's responsibility for quality is confined to "monitoring the quality of care provided under the auspices of the institution." Five years later, in the document "Guidelines on Physician Involvement in Governance of Health Care Institutions," the AHA (1987) was more specific about the board's responsibility for quality:

> A primary responsibility of a hospital governing board is to achieve and maintain high-quality patient care. Board's have as much responsibility for the quality of the health services delivered as they have for an institution's financial affairs. (Emphasis added.)

The relative importance of the board's roles and responsibilities are subject to change. But the core functions and ultimate responsibilities of the active/effective hospital board are now generally regarded as being consistent with those expressed in the AHA guidelines with the additions of a greater emphasis on quality and on board self-evaluation.

APPENDIX B

Current Characteristics of Hospital Boards: Structure, Composition, Relationships

In the summer of 1985, the American Hospital Association's Division of Hospital Governance and the AHA's Hospital Research and Educational Trust designed and conducted a comprehensive survey to assess the characteristics of hospital governance. Survey forms were sent to all 5,800 U.S. community hospitals and each hospital CEO was requested to complete the survey in conjunction with the chairman or other key officer of the board. A total of 3,189 responses, 55 percent of all acute-care U.S. community hospitals, were received. The following information is presented from the survey results (Alexander, HRET, 1986).

1. *Board Size.*

   In 1985, the average size of American hospital boards was 14 members most of whom had full voting privileges. Non-profit hospital boards averaged 18.5 members, religious hospital boards averaged 15.9 members, for-profit hospital boards averaged 9.2 members, and government hospital boards averaged 7.6 members.

2. *Age of Board Members.*

The majority of for-profit hospital board members, 60.5 percent, were between the ages of 31–50, with 37.7 percent of them between 51–70 years of age. This group of trustees was, on average, younger than the trustees of other types of hospitals. Only 36.2 percent of non-profit hospital trustees were between the ages of 31–50 years; 55.8 percent of them were between 51–70 years of age; and 7.4 percent of them were 71 years of age or older. Further, 54.6 percent of religious hospital trustees were between 51–70 years of age; 40.4 percent were between 31–50; and 4.6 percent were 71 or older. Interestingly, 1.3 percent of government hospital boards were composed of individuals younger than 31 years of age, a percentage of young trustees more than twice that of any other category of hospital board.

3. *Average Composition of the Hospital Board by Occupation.*

About 46 percent of the average board's membership had business or financial backgrounds (one possible reason to explain the tendency of many boards to define their primary priority as finance). About 30 percent of boards were composed of trustees with professions or backgrounds in health care, such as physicians, nurses, the hospital CEO, auxillians, and other health professionals. The third largest category of trustees classified by background or occupation was 16.3 percent categorized as "other" which included homemakers, farmers, and educators. Also, 5.2 percent of the average board's membership practiced religious professions, and 3 percent had government or labor positions.

This membership profile of hospital boards was largely created by the criteria used to select trustees. The survey found that the top three criteria for choosing hospital trustees, in order of frequency, were: financial and business skills, community involvement, and political influence in the community.

4. *Terms of Office.*

It is worth noting that only 41 percent of hospitals surveyed limited the number of years a board member could serve on the board. The average maximum term varied by board position with the average maximum board chair's term being 4.77 years and the average maximum term of at-large members of the board being 6.64 years.

5. *Hospital Board Relationships with the CEO.*

Eighty-seven percent of responding hospitals reported that the hospital CEO was a member of the board. Of these, only 36 percent had full-voting privileges on the board, and the remaining 51 percent of CEOs were Ex-Officio, nonvoting members of their hospital board. In about 13 percent of the responding hospitals the CEO was only an observer at board meetings; and in a small but notable situation, only .4 percent of hospital CEOs did not attend the meetings of their hospital board.

Of responding hospitals, only 42 percent of hospital boards had written employment contracts with their CEOs. Of those CEOs with employment contracts, the average length of the contract was three years. The most frequent provisions in a CEO's contract related to job responsibilities, compensation, and termination provisions. The least frequent provisions were CEO performance standards and conflict of interest provisions.

6. *Hospital Board Relationships with the Medical Staff.*

Physician involvement in the governance of the hospital is one method for integrating physicians into the hospital decision-making structure and process and for facilitating good board relations with the medical staff. Approximately, 77 percent of responding hospitals reported having at least one physician on the governing board. Thirty-three percent of the physician trustees had medical staff privileges at the hospital at which they were a governing board member. Of all physician trustees, 35 percent had full voting privileges on their board, and 31 percent had both full board voting privileges and medical staff privileges at the hospital.

Another method of facilitating good communication between the board and medical staff is by having trustees serve on medical staff committees. Interestingly, the most frequently cited medical staff committee with trustee membership was the medical staff quality assurance committee, with 35 percent of hospitals reporting this. Twenty-two percent of hosptials reported nonphysician trustee membership on the medical staff committee, and 18 percent reported trustee membership on both the medical staff credentials and utilization review committees.

7. *Board Self-Evaluation and Continuing Education*

Despite the then just publicized Joint Commission standard requiring board self-evaluation, in 1985 only 22 percent of hospital boards had formal evaluation mechanisms. Of those hospitals that did conduct self-evaluations, 75 percent of them did so annually. Regarding trustee continuing education, only 25 percent of hospitals had written policies requiring or relating to board education, but 35 percent of hospitals reported allocating budget funds for continuing education of trustees.

Index

A

Accreditation Manual for Hospitals, 26
Action plan
  advanced, 108–9
  basics of, 97–103
  developing, 96–97
  evaluation, 107
  sample, 103–9
Active/effective hospital board, 9, 11, 12, 80, 81, 97
  characteristics of, 16–17
Adverse patient occurrences, 141
Agenda issues, 31–32
Annual reappraisal, 58, 141
Anthony and Singer, 28
Antitrust, 43
Applicant capability, competency, 63
Appointments, medical staffing, 61–69, 141
Arithmetic mean, 139–40
Assessment, 96
  checklist, 81–91

B

Background information, 116
Bader, Barry, 34–35
Bing v. Thunig, 23, 25, 29–30, 43
Board Quality Assurance Committee, 1
Board structure, 149–57

C

Calendar, 143–44, 146
  sample, 144–45
Capability, 63
Chief Quality Officer, 3
Clinical information, 60–61, 111–13
Clinical norms, 132
Communications, 7–8, 41–42
Competency, 63
Concurrent review, 47–48
Consultant, 37–38
Continuous-improvement approach, 49, 53–55
Corleto v. Shore Memorial Hospital, 29

Credentialing, 62–63, 68–69, 87–89
  criteria levels, 68–69
Cross-sectional data, 115–16
Custodially-related patient injuries, 72–73

D

Darling v. Charleston Community Memorial Hospital, 10, 24–25, 30, 33, 43
Data, 114
Dealbreaker criteria, 68
Deaths, graphs, 134–35. *See also* Mortality graphs.
Decision criteria, 67–68
Denominator data, 114
Disciplinary action, 36

E

Education, 82–83, 99–100, 103–4
  emphasis, 9
Educational reports, 92
Educator, 37–38
Effective hospital board. *See* Active/effective hospital board.
Effectiveness emphasis, 9
Evaluation, 100, 147
  emphasis, 9
  process, 53–55, 67
Evaluation process, 53–55
External information, 75–76

F

Fiduciary duty, 43–44
Financial viability, 32–33
Franklin, Benjamin, 8–9
Fundraising emphasis, 8–10

G

Goals, 95, 100, 132
Governance information, 111–13, 140
Governing board
  role of, 64
  standards, 26–27
  See also Hospital board.
Greater Southeast Community Hospital Foundation, 119
Guidelines for board information, 124–27

H

Hammurabi Code, 19
Health Care Financing Administration, 75, 142
Health Care Quality Improvement Act of 1985, 36, 75
Hippocrates Oath, 19
Honorific hospital board, 9–10, 11, 12, 80, 81, 97, 101
  characteristics of, 15
Hospital board
  assessing current role in QA, 80–93
  assessing its functioning mode, 13–17
  credentialing oversight, 87–89
  QA committee, 52–53
  quality oversight, 84–86, 105–6
  role in quality, 1–2, 13, 19–25, 79–80, 83–84, 104–5
  role statement for, 122–24
  stages of, 9–11
  structure and quality, 89
  why responsible for quality, 42–44

Index 171

Hospital governance
  board's general mode of functioning, 13–17
  board roles and duties, 13
  development and evolution of, 7–12, 17–18
Hospital licensing statute, 24
Hospital Research and Educational Trust agenda issues, 31–32
Hospitals, 32
Hospitalwide nosocomial infections, 136
  with threshold level, 136
Hospitalwide deaths, 133–34
  with threshold level, 135
Hospitalwide plan, 59

I

Iatrogenic patient injuries, 72–73
Indicators, quality, 127–30
Infection control, 69–71, 142
Infection Control Committee, 71
Information, 55–56
  defining other quality-related, 140–43
  external quality-related, 75–76
  flow, 86–87
  governance vs. management and clinical, 111–13
  guidelines, 124–27
  patient satisfaction, 74
  process for developing QA, 121–22
  QA reports to board, 113–21
  quality flow and reporting, 106
Initial appointments, 64–65

J

Johnson v. Misericordia Community Hospital, 29, 43
Joint Commission on Accreditation of Healthcare Organizations (JCAHO), 51
  annual QA reappraisal, 58
  Governing Body standards, 26–27, 43, 44, 92
  monitoring and evaluation process, 53–54
  scoring guidelines, 27–28
  surveys, 75, 94, 142

K

Keys to Better Hospital Governance Through Better Information, 34–35
Korn/Ferry 1989 survey, 32

L

Legal relationships, 20–21, 25. *See also specific legal cases.*
Level 1 criteria, 68
Level 2 criteria, 68–69
Licensure survey reports, 142
Longitudinal information, 115

M

Malpractice, 29, 75, 142
Management information, 111–13
Medicaid, 10
Medical staff
  appointments, 61–69
  credentials, 22

privileges, 63
  role of, 63-64
Medical Staff Executive Committee, 62
Medicare, 10
  Conditions of Participation, 28, 43
  mortality data, 142
Meeting minutes, 116-17
Michigan Legislature's 1968 hospital licensing statute, 24
Michigan Supreme Court's 1902 decision, 22
Monitoring, 53-55
Mortality
  graphs, 134-35
  department of surgery graphs, 138-39

N

National averages, published, 132
National Practitioner Data Bank, 36, 142
New Jersey Supreme Court 1975 ruling, 28-29
New York Court of Appeals' 1957 ruling, 23
New York State of Court Appeals 1914 ruling, 22
Nosocomial infections, 69-71
  graphs, 136-37
Numerator data, 114-16

O

Orientation, 82-83, 99, 103-4
Outcome, 46, 118
Over-involvement, 35

P

Passet, Barry A., 119
Patient satisfaction information, 74
Patrick v. Burget, 43
Pennsylvania Hospital, 8
Pepke v. Grace, 22, 24
Performance-based review, 65-66
Philanthropic hospital board, 9-10, 11, 12, 80, 81, 97, 101
  characteristics of, 15
Physician incompetency, 29
Plan, 57-58
Presentation formats, 119
Privileges, 141
Problem-focused approach, 49, 52-53
Process, 46
Prospective quality assurance, 48, 98, 108

Q

Quality
  board structure and, 89, 106-7, 149-57
  defined, 44-45
  in general, 41-42
  inhibiting factors, 35-36
  measured, 45-49
  -related information, 140-43
  why boards are responsible for, 42-44
Quality assurance
  clinical data, 60-61
  committee, 151-52
  defined, 49-50
  information, process for developing, 118, 121-22

monitoring, evaluation
    process, 53–55
plan, 57–58
problem-focused, 52–53
process, final steps, 146–47
program information, 155–60
and quality, 55–56
reports, typical weaknesses
    of, 113–21
review, 57
why do, 50–52
Quality assurance professionals
    and board involvement,
    93–95
future of, 155–57
role of, 64
role in board oversight, 3–5,
    36–38
Quality indicators, 127–30
report formats for, 133–40
threshold levels for, 130–33
Quality information flow and
    reporting, 106
Quality of care
    current hospital board role
        in, 26–30
    historical role of hospital
        board in, 19–25
    improvement of, 59–60
    information, 55–56
    oversight of, 22–23
    role of QA professional in,
        36–38
    why hospital boards are
        uncomfortable, 30–36

R

Readmission, 63
Reappointments, 64–65
Reappraisal, 58
Refinements, 147
Reporting calendar, 143–46

Reports, 37
    common flaws of, 113–21
    formats, 133–40
Retrospective review, 47
Review, 57–61
Risk management, 71–73,
    141–42
Role identification, 57
Rubber stamping, 62, 66
RUMBA, 140

S

Salinsky, Dr., 29
*Schloendorff v. Society of New
    York Hospital,* 22
Self evaluation, 92
Staffing patterns, plans, 74–75
Statement of board's role,
    122–24
    sample, 123–24
Structure, 46
    improvement options, 150–51
    introduction to, 149–50
Summary information, 61
Surgery mortality rates, 138–39

T

Target/threshold level, 130–39
Time frames, 100, 103–7
Transformation, 95
Transition hospital board, 9,
    10–11, 81, 97
    characteristics of, 16
    sample action plan for, 102–7

U

Under-involvement, 35
University of Minnesota at
    Minneapolis 1989 survey,
    32

W

Wisconsin Court of Appeals
1981 ruling, 29

X

X-ray waiting period, 137, 139